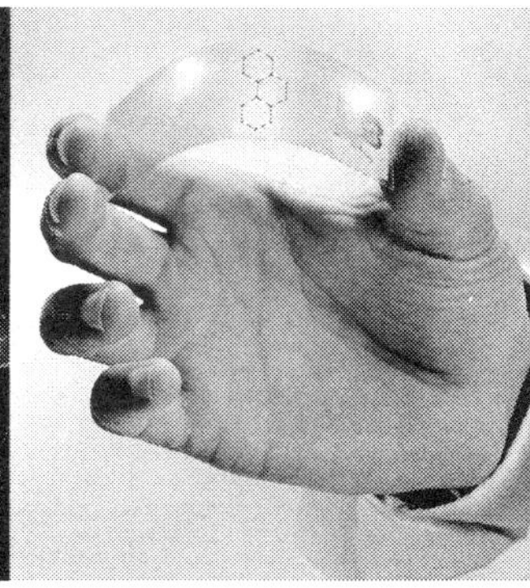

Graphene:
charge transport and flexible nanoelectronics

Tae-Jun Ha Jong-Ho Lee

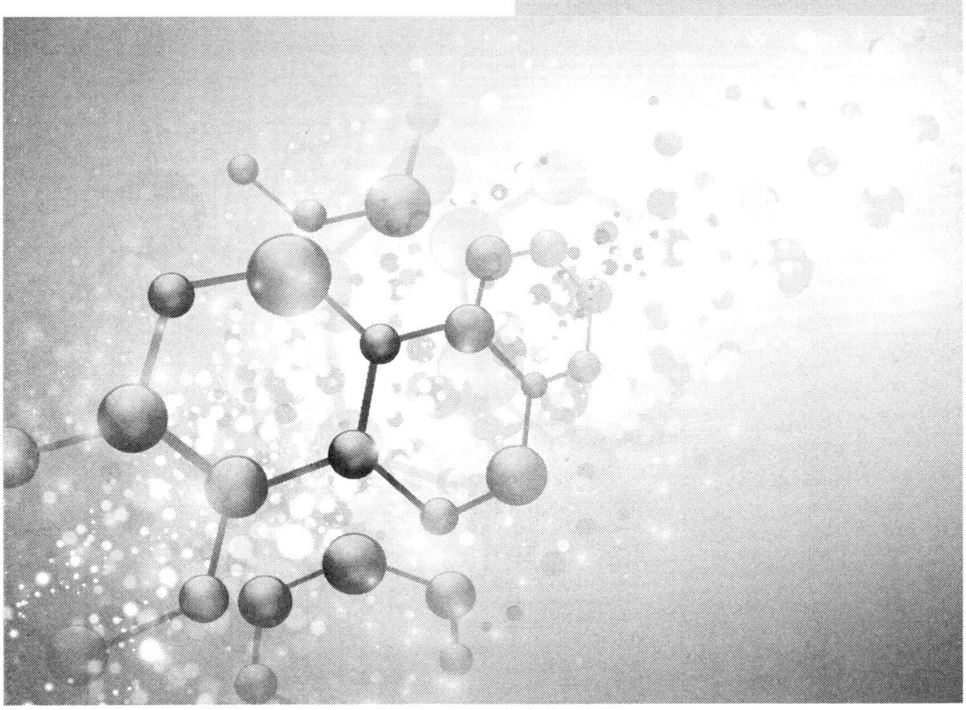

도서출판 동화기술

Graphene: charge transport and flexible nanoelectronics

Despite the widespread interest in graphene-based electronics over the last decade, high-performance graphene field-effect transistors (GFETs) on flexible substrates have been rarely achieved even though this atomic sheet is widely understood to have greater prospects for flexible electronic systems. Since the key device metrics of GFETs such as carrier mobility, Dirac voltage, impurity concentration and on-off current ratio are affected by charge transport, comprehensive understanding on the device physics is important to realize high performance electronics. In this book, we discuss the charge transport in GFETs and demonstrate the realization of high-performance GFETs implemented on flexible plastic substrates. The structural integration of materials between graphene and organic/polymer is explored to understand device physics underlying the charge transport in GFETs. The optimum device structure for high charge velocity and bendability is suggested with experimental comparison among diverse structures including top-gate GFETs (TG-GFETs), single/multi-finger embedded-gate GFETs with high-k dielectrics (EG-highk/GFETs), and embedded-gate GFETs with hexagonal boron nitride (h-BN) dielectrics. Flexible graphene transistors with high-k dielectric afforded intrinsic gain, maximum carrier mobility of 8,000 $cm^2/V \cdot s$, and importantly 32 GHz cut-off frequency. Mechanical studies reveal robust transistor performance under repeated bending down to 0.7 mm bending radius whose tensile strain corresponds to 8.6%. Passivation techniques, with robust mechanical and chemical protection in order to operate under harsh environments, for embedded-gate structures are also covered. The integration of functional coatings such as highly hydrophobic fluoropolymers combined with the self-passivation properties of the polyimide substrate provides water-resistant protection without compromising flexibility, which is an important advancement for the realization of future robust flexible systems based on graphene.

Table of Contents

5	Chapter 1	Introduction
15	Chapter 2	Preparation of Electronic-Grade Flexible Substrates
23	Chapter 3	Transistor Device Topology
33	Chapter 4	Design Considerations for Optimization
49	Chapter 5	Experimental Results : Electrical / Mechanical Performances
69	Chapter 6	Transformation of the Electrical Characteristics in GFETs
95	Chapter 7	Summary
97	Chapter 8	Appendix
113		References

Chap 1

Introduction

- Carbon-based materials
- Electronic structure of graphene
- Electronic properties of graphene
- Charge transport in graphene
- Graphene Field-Effect Transistor
- Flexible Electronics

Carbon-based materials

Figure 1.1 shows the allotropes of carbon from 0-dimensional to 3-dimensional structures [1]. One example of a 0 D allotrope is C_{60} which has a soccer ball shape and consist of 60carbon atoms, 1 D forms include carbon nanotubes (CNTs), the 2 D form isdesignated graphene and 3 D forms include diamond (net structure) and graphite (sheet structure). As shown in Figure 1.2, it is possible to generate many of these allotropes from a sheet of graphene [2].

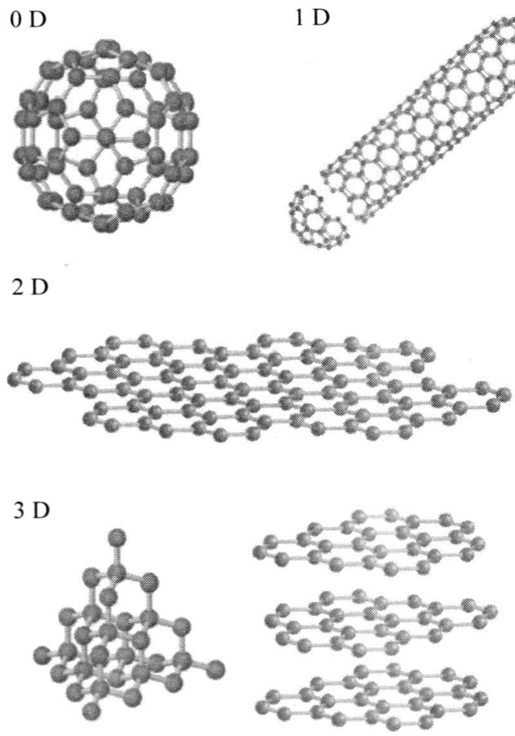

Figure 1.1 The allotropes of carbon from 0 dimensional to 3 dimensional structures [1]

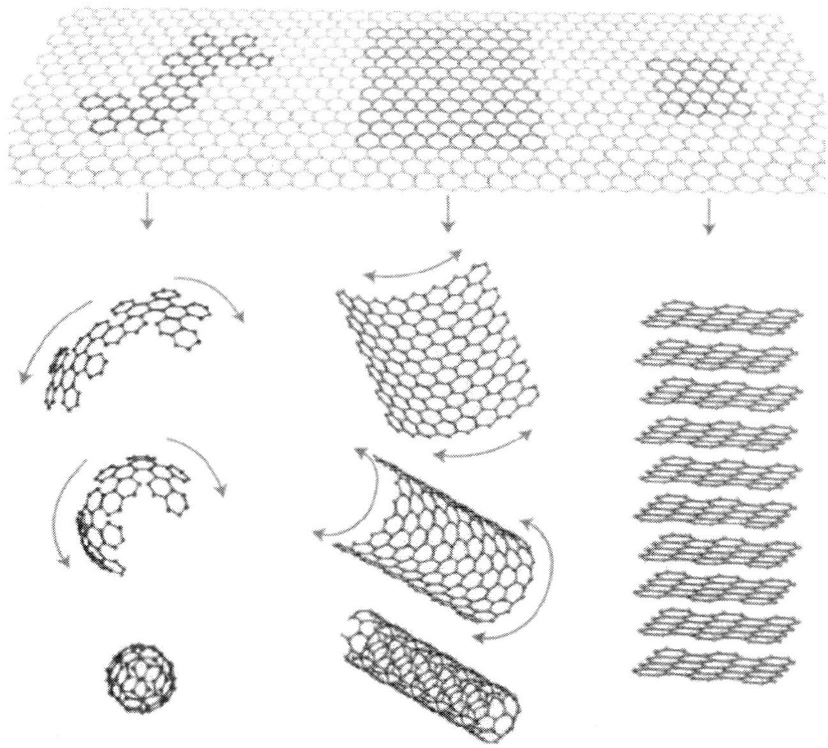

Figure 1.2 The illustration of how to make different dimensional forms from graphene [2]

Electronic structure of graphene

Graphene (defined as being a single layer of graphite) is a two-dimensional hexagonal lattice consisting of carbon atoms [2-3]. Each carbon atom is covalently bonded with nearest neighbors through sp^2 hybridization. The p bands are well-characterized as linear combinations of porbitals of the carbon atoms [4]. The unit cell is based on two carbon atoms (two nonequivalent sub-lattices, A and B) and the carbon-carbon bond length is 1.42 Å [4]. Hopping transport between the sub-lattices results in the formation of two energy bands, and the structure is established with two conical points per Brillouin zone where band crossing occurs at the K and K' points. These points are called Dirac point as shown in Figure 1.9 [3]. The band structure of graphene as shown in Figure 1.3, indicates that the conduction band and the valence band intersect at the Dirac point and the electron energy is linearly dependent on the wave vector. As a result, quasiparticles in graphene exhibit a linear dispersion relation $E = hk\upsilon F$, as if they are massless relativistic particles [5-6]. In graphene, the charge carriers mimic particles, which are governed by the Dirac equation rather than the Schrödinger equation which is commonly employed in the analysis of conventional solid state systems where effective mass corresponds to the curvature of the band structure [6]. Based on the band-structure, intrinsic graphene is a semi-metal or zero-gap semiconductor [7].

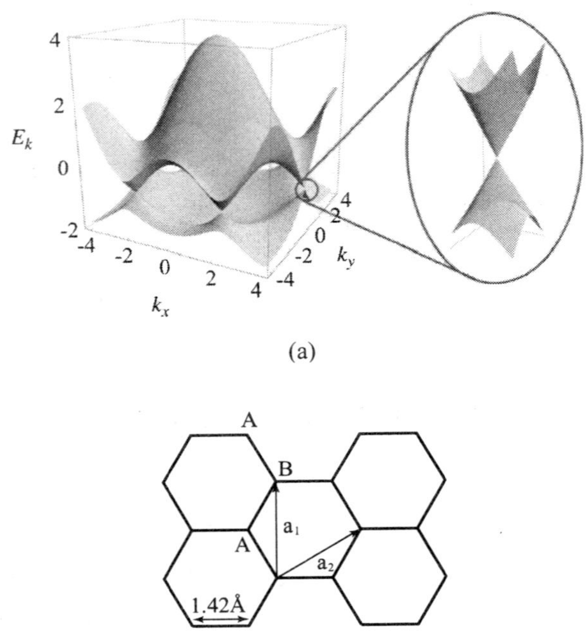

Figure 1.3 (a) Band structure of graphene [3]
(b) sub-lattices structure

Electronic properties of graphene

It was reported that the electron mobility at room temperature in exfoliated graphene is excess of 15000 cm^2/V-s at a carrier concentration of 10^{12} cm^{-2} [8]. In addition, suspended graphene exhibits the mobility of up to 200000 cm^2/V-s with carrier concentration of 2×10^{11} cm^{-2} at room temperature [9]. However, since scattering of electrons by optical phonons from the substrate dominates at room temperature, device performance is limited in graphene FETs [10-11]. Furthermore, impurities that incorporate into the film during the fabrication process also become a scattering source. Recently, boron-nitride (BN) has been used as a substrate for the graphene since BN possesses a high phonon energy and does not incorporate impurities due to the absence of out-of-plane bonds [12-13]. Figure 1.4 shows the scattering mechanisms in graphene [14]. Whereas surface optical phonon scattering is dominant at lower electron energy, optical phonon scattering limits the mobility at higher electron energy.

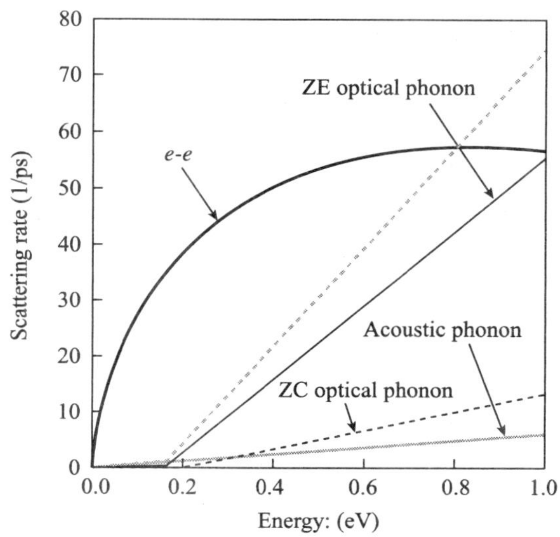

Figure 1.4 Scattering rates vs. energy in graphene [14]

In addition, since graphene has a zero band-gap, FET devices possess high off-currents, making it difficult to use them as logic devices, which require on-off current ratios in excess of $10^4 - 10^6$ [15]. Graphene must have a band-gap energy of at least 0.4 eV to be useful in logic device applications [15]. Instead of applications that demand a large on-off current ratio, graphene FETs are expected to be useful in radio frequency (RF) applications. There have been previous attempts to improve the on-off current ratio in graphene FETs [16-17]. In addition, graphene FETs with a 100 GHz cut-off frequency have been achieved, which is a remarkable result since it is higher than what silicon FETs have been able to achieve [18]. However, absence of current saturation in the output characteristics of graphene is not ideal for RF device performance.

Charge transport in graphene

(1) Diffusive transport

Chen et al. and Hwang et al. reported that the transport in graphene is affected by charged impurities which act as scattering centers through long range Coulomb scattering [19-20]. Figure 1.5 shows graphene conductivity limited by Coulomb scattering. The linear relation between conductivity and carrier concentration is observed.

(2) Self-consistent theory

The charged impurities present random voltage fluctuations in the graphene layer, which leads to a shift in the Dirac point and induces a residual carrier density due to electron and hole puddle formation caused by the charged impurities [21]. The charged impurities act as scattering centers and also reduce the mobility. Figure 1.6 shows the results of self-consistent calculations [21]. The results indicate that the graphene minimum conductivity is not universal and the shift in Dirac point is related with the residual carrier density induced by potential fluctuation.

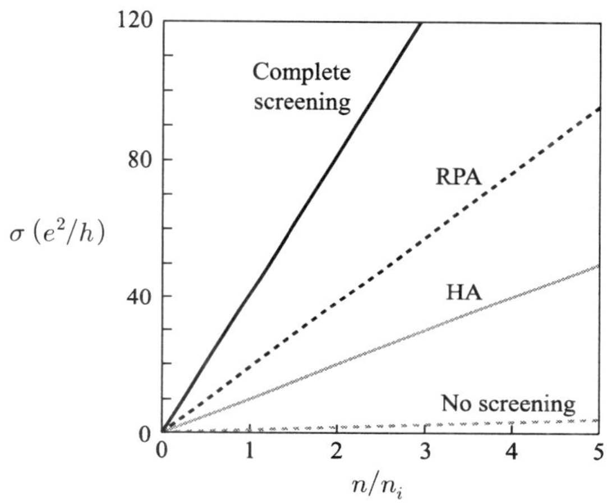

Figure 1.5 Graphene conductivity as a function of carrier density with random phase approximation (RPA) and Hubbard approximation (HA) screens [20]

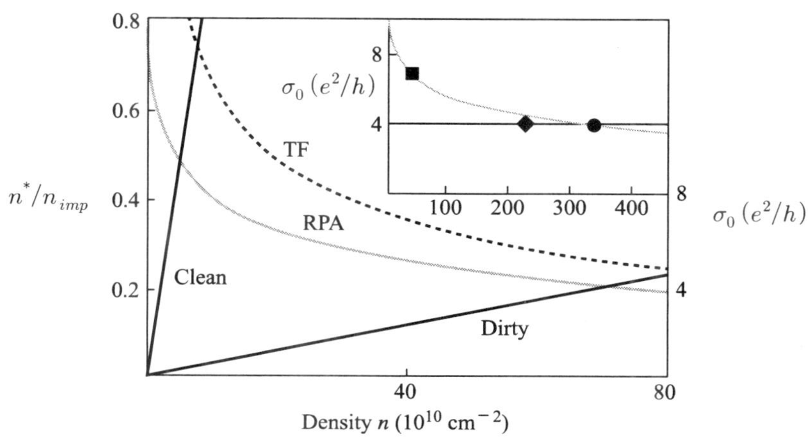

Figure 1.6 RPA and Thomas-Fermi (TF) as a function of carrier density: The points of intersection represent the self-consistent solution [21]

(3) Ballistic transport

The charged impurities reduce the mean free path of charge carriers in graphene and make the transport diffusive. If this effect is ameliorated, graphene devices can possess very fast charge ballistic transport [22-23]. $4e^2/\pi h$ in the minimum conductivity at the Dirac point is the theoretically predicted value for ballistic transport.

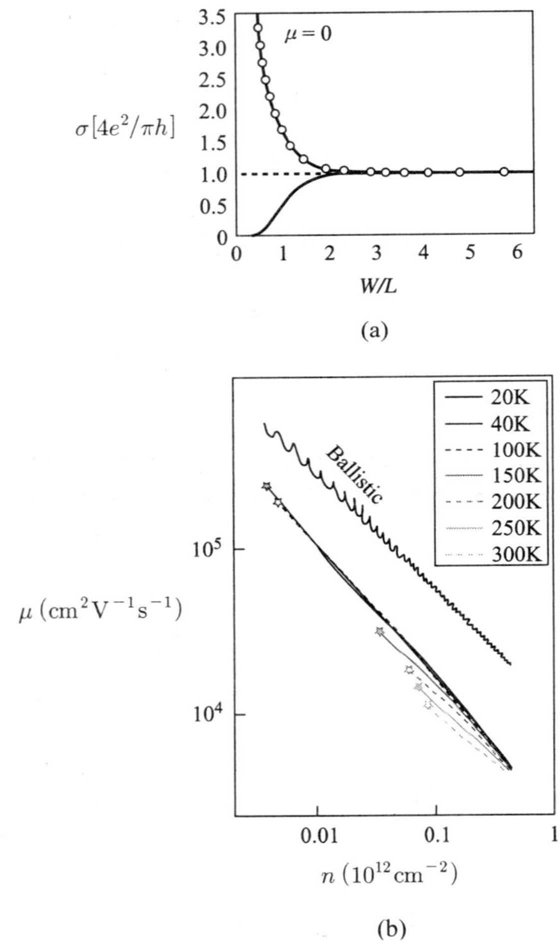

Figure 1.7 (a) Conductivity at the Dirac point as a function of the ratio of graphene strip and [22]
(b) Carrier density dependence on the mobility with different temperatures compared with Ballistic model [23]

Graphene Field-Effect Transistor

The outstanding charge transport in graphene field effect transistors offer attractive prospects for high-speed analog and radio-frequency (RF) electronics. These outstanding features including high mobilities at room temperature, sub-THz cutoff frequencies at moderate channel lengths, and ambipolar electron-hole symmetry, afford linear and non-linear analog signal processing with the simplicity of a single transistor [24-28]. In addition, graphene is substrate agnostic and mechanically flexible. Hence, it is an ideal material for flexible electronics on polymeric substrates. Although intrinsic carrier mobility of graphene approaching 100,000 cm^2/V-s has been reported at room temperature on suspended devices, the carrier mobility of graphene degrades on oxide supported graphene field-effect transistors (GFETs) with values around 10,000 cm^2/V-s frequently measured on clean exfoliated graphene on SiO_2 in agreement with theoretical upper limits [29-32] However, it is unclear how fast charge carriers can travel in monolayer graphene on polymeric substrates, which is an important substrate for flexible electronics. It has been suggested that remote phonon scattering from the broad continuum of modes from the polymer substrate should impact the carrier mobilities at a scale comparable to SiO_2 [33]. As graphene holds great potential for fast flexible electronics, it is crucial to experimentally access its fast carrier transport on polymer supports.

Flexible Electronics

The field of flexible electronics has been active for more than a couple of decades, driven by the desire for the realization of low-cost, large-area electronics. The development of flexible electronics operating at radio-frequencies (RF) requires materials that combine excellent electronic performances and the ability to withstand high levels of strain. Circuit components required to implement flexible devices include (1) a supporting flexible film as a substrate, (2) electrically conducting films as electrodes, (3) dielectric films to be used as gate dielectrics and device isolation/passivation layers, and (4) active channel materials. The technical bottleneck in this field of research has been the preparation of high performance active channels due to the process compatibility to flexible substrates. Due to limited carrier mobilities of less than 100 cm^2/V-s for diverse semiconducting materials reported as flexible channels, the resulting electronic device performance has been relatively poor [34-36]. The desire to improve the performance of these devices has led to the introduction of other materials which include carbon nanotubes, silicon thin-films and nanowires, and compound semiconductors [37-42]. III-V compound semiconductors and graphene show the highest mobility up to 10,000 cm^2/V-s on plastics while others show still relatively low values. However, enhancements in electronic performance have been achieved at the expense of device flexibility. In addition to the high carrier mobility and GHz cut-off-frequency, graphene is able to survive the mechanical tensile strain above 20%, and offers a technical method for wafer-scalability [44-46].

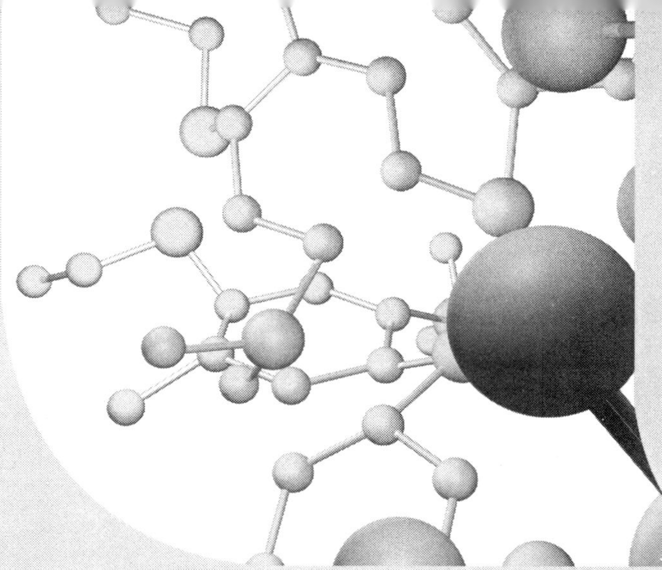

Chap 2

Preparation of Electronic-Grade Flexible Substrates

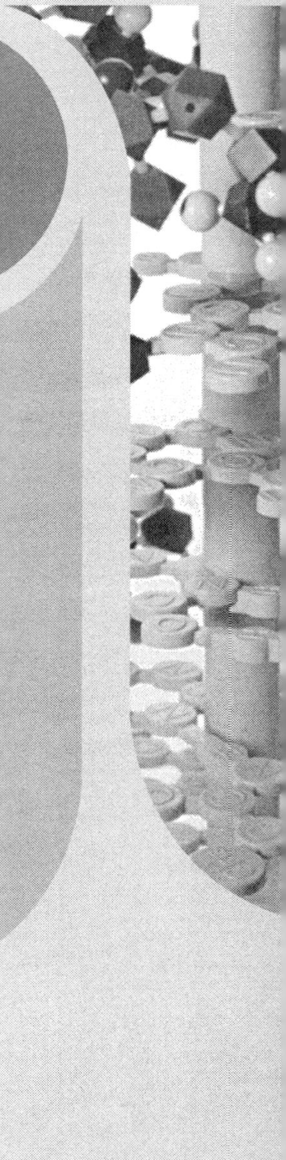

- Requirements for Flexible Substrates
- Commercially Available Flexible Substrates
- Surface Preparation of Polymer Substrates for Device Fabrication

Requirements for Flexible Substrates

Substrates to be used as a flexible supporting layer replacing the conventional rigid substrates should meet several requirements. Those requirements are listed as follows.

1. Surface quality / roughnessHighly uneven surface topography limiting the fine resolution of lithographic steps has to be avoided. Nanometer-scale short/long range roughness values are desirable to prepare sub-micron channel length devices. Roughness over long distance (or more generally thickness variation) larger than the unit device structure is still acceptable. Chemical-Mechanical-Planarization (CMP) or other surface treatment processes are usually required for commercial flexible substrates to achieve acceptable roughness values.

2. Thermal propertiesThe maximum processing temperature for the flexible substrates should not exceed their glass transition temperatures (Tg). During the exposure to high temperature processes even with highest temperature less than Tg, the mismatch in thermal expansion coefficients of integrated films and its resulting built-in stress of the sample have to be minimized. Metallic foils and glass substrates are superior in this aspect than polymer substrates.

3. Mechanical propertiesMaterials with high elastic modulus are less flexible and easier to handle during the fabrication. Highly elastic polymers and elastomers may require hard carrier wafers, where flexible substrates are attached, during the device fabrication.

4. Chemical properties/compatibilitiesSubstrates should not react with chemicals (acids and solvents) used during device fabrication. Different coating layers may be required to protect the underlying base substrates.

5. Electrical conductivityMetallic foils are electrically conductive and require additional electrical isolation. Glass substrates and most polymers are good insulators.

6. (Optional) Optical clarityNot always required, but becoming an important metric for transparent electronics or optoelectronic applications.

Commercially Available Flexible Substrates

(1) Glass substrate

Even though glass plates are considered to be hard and rigid, they become flexible when their film thickness are thinned down to less than 100μm [47-48]. Glass substrates have high optical transparency and thermal stabilities (low thermal expansion coefficient and high glass transition temperature) suggesting good opportunities for wide-range of electronic applications including flat-panel displays. Mechanical and chemical properties are also attractive. However, they are brittle and cannot tolerate high level of strain, which can be introduced while handling them for fabrication.

(2) Metallic foil

Stainless steel is one of the popular metallic foils currently being considered as a flexible substrate. They also become flexible similar to glass substrates when their film thickness are thinned down to 100μm-scale, though their surface qualities are lower than that of glass substrates [49-50]. Metal foils are not transparent and electrically conductive, however can be still attractive for such applications where reflective surfaces are required. Surface planarization layers are also applied to offer electrical isolation and/or further improve the surface quality [51-52]. Inorganic materials such as plasma-enhanced chemical vapor deposited (PECVD) silicon nitride and oxide are also used to give the electrical isolation.

(3) Polymer substrate

Polymers offer great opportunities to be used as flexible substrates. Diverse candidates are commercially available and a part of them are listed in Table 2.1. These are highly flexible and inexpensive. However, thermal properties are a major concern while utilizing these materials for electronics. Low thermal conductivities along with low glass transition temperatures of polymers limit several processing, and operating conditions.[53-55] Dimensional stabilities are also not as great as other candidate substrates listed above. Though high flexibility and low elastic modulus are among desirable features here, when these candidates are exposed to thermal processes repeatedly during device fabrication, the resulting built-in thermal stresses induce non-negligible

Table 2.1 Material properties of commercial flexible substrates

Property	Unit	Flexible substrates				
		Glass	Stainless steel	Polymer		
				PEN	PET	PI
Transparency		Good	N/A	Good	Good	Poor (Orange)
Max. process temperature	°C	600	900	155	120	300
CTE	ppm/°C	3~4	18	20	20~80	15~35
Elastic modulus	GPa	70	200	5	2~4	2~3
Electrical conductivity		Low	High	Low	Low	Low
Thermal conductivity	W/m°C	1	16	0.15	0.15~0.4	0.1~0.35

film curvatures due to mismatch in their thermal expansion coefficients. Material properties of commonly used flexible substrates are listed in Table 2.1 [56-60].

Surface Preparation of Polymer Substrates for Device Fabrication

Flexible industrial polyimide sheet (50~100μm thick) was chosen as the substrate because of its high glass transition temperature (>300 ℃), Young's modulus (~3.5 GPa), and high solvent resistance making it compatible with conventional device fabrication. The polyimide sheets used in this study are commercially available from American Durafilm Inc. (#300 FPC) and Dupont Inc. (Kapton film). Industrial polyimide (PI) has an uneven topography with surface roughness that can exceed 100nm [61]. To make the un-even surface of the industrial PI films smooth, liquid polyimide (PI 2574 from HD Microsystems) was spin-coated onto the polyimide sheet. The coated substrate was soft-baked at 200 ℃ for 30 min and subsequently cured at 300 ℃ for 1 hour under a nitrogen atmosphere to obtain a smooth surface (RMS ~ 1~2 nm), necessary for preserving the quality of electronic materials and for achieving the best results in lithographic resolution and registration [61]. This coated polyimide sheet was then carefully mounted onto a supporting Si carrier wafer to ensure the flatness for the following fabrication process. Prior to use, the polyimide sheet was rinsed again in Acetone and 2-propanol (IPA) for 5 min each. The surface roughness of the polyimide surface before and after coating was measured using a VeecoVR D-5000 atomic force microscope (AFM) in tapping mode. Figure 2.1 shows the AFM images of polyimide surfaces scanned over the as-received commercial PI film and the PI film after the surface treatment. To further improve the surface roughness, additional step of depositing high-k atomic layer deposition (ALD) dielectrics is done to improve the surface. This ALD dielectric layer can work as i) the high quality gate dielectric for embedded-gate device structures, and ii) the heat spread layer, which is an important component for the thermal management on flexible substrates, which will be discussed in a later chapter.

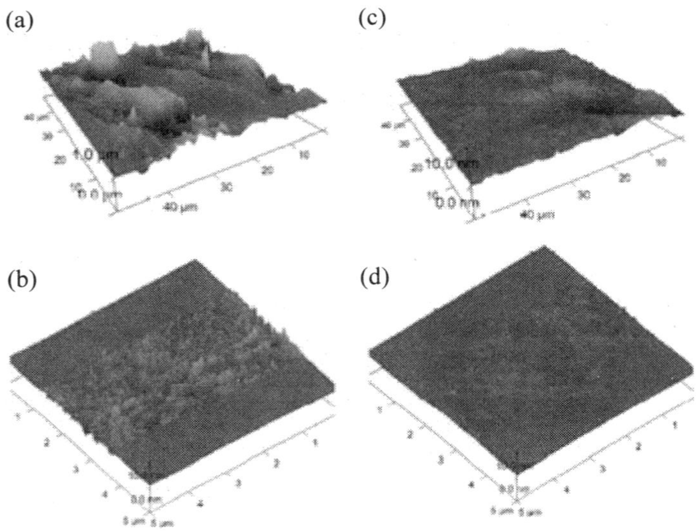

Figure 2.1. AFM images of polyimide surfaces.

(a) $50 \times 0 \mu m^2$
(b) $5 \times 2.5 \mu m^2$ scan on polyimide sheet as received
(c) $50 \times 50 \mu m^2$
(d) $5 \times 2.5 \mu m^2$ scan after surface treatment

Despite the attractive features of graphene, diverse charge scattering mechanisms prevent access to the intrinsic electrical performance. These challenges include ionized impurity scattering which degrades the mean-free-path, substrate and interface charge traps resulting in charge puddles, and random shifting of the charge-neutrality point [62-64]. Each of the mechanical and chemical non-idealities can result in deviation of the band structure and transport characteristics, emphasizing the need for a high purity, ultra-smooth dielectric that is compatible with graphene devices. Hexagonal boron nitride (h-BN) is an insulating isomorph of graphene with boron and nitrogen arranged in a hexagonal lattice, with a small lattice mismatch of ~1.8% compared to graphene [65]. Several features of h-BN make it an excellent candidate dielectric for GFETs on flexible substrates. The ultra-flat surface of h-BN significantly reduces the electron–hole charge fluctuations, compared to polymeric interfaces or SiO_2. This enables the observation of near-ideal graphene properties without the use of complex suspended structures [65]. Owing to its strong in-plane covalent bonds and relatively weak inter-plane bonds due van der Waals interactions, h-BN is cleaved and exfoliated in the same way as graphene [65-66].Recently a series of reports have suggested that graphene/h-BN heterostructures enhance field-effect performance [66-67]. Here, we report the method of utilizing an h-BN film as a gate dielectric for flexible FETs. The preparation of the substrate is as follows. h-BN (from Momentive Performance Materials) was exfoliated on silicon wafers with 285-nm thick thermal oxide, which offers good optical contrast between thin h-BN films and substrates for easy identification. The sample was first annealed under nitrogen at 300 °C for 1 hour to achieve a smooth h-BN surface and remove residual adhesive on its surface to define a better interface between gate electrodes and h-BN films. Gate electrodes can be also patterned over annealed h-BN films on SiO_2/Si substrates by electron-beam lithography (EBL), e-beam evaporation, and liftoff to prepare the embedded-gate structures. Liquid-type PI (PI-2574 from HD Microsystems) was spin-coated on the patterned sample and cured under nitrogen to capture the device structures within PI. Thermal oxide was then etched by buffered-oxide-etchant (BOE) and the PI films with embedded gates and dielectrics were released (capture-release process). The released sample was then annealed again under nitrogen to provide a clean high-quality interface between graphene and dielectrics (two-step annealing process). Figure 2.2 shows the comparison between different substrate preparation techniques.

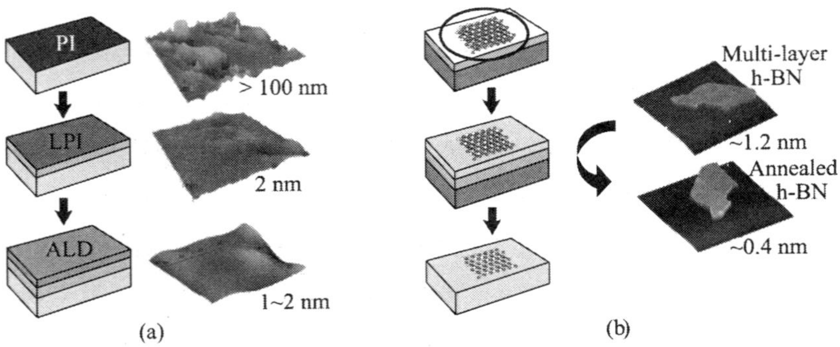

Figure 2.2. 3-D illustrations of different flexible substrates preparation techniques.

(a) The surface treatments on flexible PI films. An as-received PI film, a cured liquid PI coated on a PI film, and an ALD deposited on a cured film are shown with their AFM images and roughness values.
(b) Preparation of an embedded h-BN film captured in a free-standing PI substrate. The simplified process flow is given with the AFM images scanned over an as-exfoliated h-BN film (roughness ~1.2 nm) and an annealed h-BN film (roughness ~0.4 nm).

Chap 3

Transistor Device Topology

- Top-Gate Graphene Field-Effect Transistors
- Embedded-Gate Graphene Field-Effect Transistors
- Modified Embedded-Gate Graphene Field-Effect Transistors
- Comparison of Device Structures

Top-Gate Graphene Field-Effect Transistors

The graphene monolayer employed in this study was prepared via low-pressure chemical vapor deposition (CVD) on copper foil as reported previously [68]. A typical Raman spectrum of graphene synthesized with the CVD procedure on copper foil is shown in Figure 3.1(a). The full width at half maximum of the 2 D peak is approximately $35cm^{-1}$, the 2 D/G ratio is ~3, and the D-peak is negligible, indicating high-quality monolayer graphene. The typical solution-based poly(methyl methacrylate) (PMMA) assisted graphene transfer from the copper foil, was employed to integrate the graphene onto the PI sheet [69]. The AFM image in Figure 3.1(b) shows the substrate surface after graphene transfer on polyimide sheet with roughness of about 2 nm in small areas but with wrinkles as high as 8 nm due to the transfer process. These wrinkles from the polymer residue are detrimental to transport and mobility [70]. An array of graphene transistors were fabricated directly on to the polyimide sheet using standard microelectronic processes and electron-beam lithography with a charge compensation layer essential for patterning insulating substrates. The simplified fabrication process is illustrated in Figure 3.2(a). In brief, the first e-beam lithography patterns the active area while isolating graphene channels from each other. Oxygen plasma is used to etch the superfluous graphene. After defining source and drain electrodes by lithography and evaporation, high-k gate dielectric is deposited by ALD. The gate dielectric is composed of evaporated 1nm titanium that oxidizes readily in air at 200 ℃ and serves as the seed layer for uniform ALD of 30 nm Al_2O_3, with estimated capacitance of ~200 nF/cm^2 (k~6.8). The gate electrode is subsequently patterned. The gate and source/drain metals are Ni (50 nm) and Ni/Au(10 nm/40 nm) respectively. An optical image of a completed device with ground-signal-ground (GSG) input (GFET gate) and output (GFET drain) pads necessary for RF measurements is shown in Figure 3.2(b). Figure 3.2(c) is a photograph of the flexible polyimide substrate with an array of integrated GFETs.

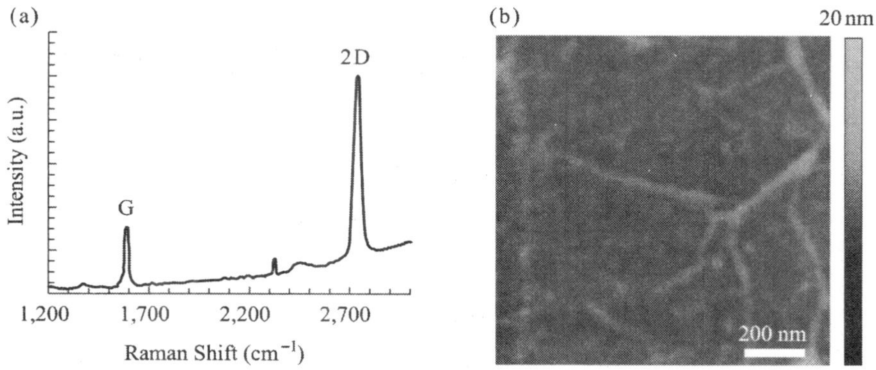

Figure 3.1. Monolayer graphene on copper foil

(a) Raman spectrum of the synthesized monolayer graphene on copper foil indicating the 2D and G peaks.
(b) AFM image of graphene transferred to flexible polyimide sheet with a surface roughness ~2 nm and wrinkles as high as 8 nm.

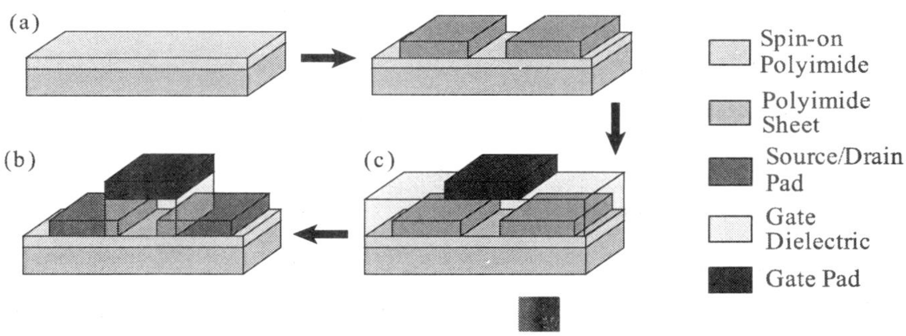

Figure 3.2. Top-gate graphene transistor on polyimide sheet

(a) Simplified cross-sectional fabrication process for graphene transistor on polyimide sheet featuring transferred CVD graphene, high-k Al_2O_3 dielectric, and metal electrodes.
(b) Optical image of a completed GFET with two gate fingers in a GSG configuration to facilitate high-frequency measurements. The device length and total width are 0.25μm and 10μm respectively.
(c) Photograph of the flexible substrate with an array of integrated GFETs.

Embedded-Gate Graphene Field-Effect Transistors

Two types of embedded-gate graphene field-effect transistors (EGFETs) have been realized in this work. The first one is a detachable EGFET. The fabrication process for the detachable EGFET is illustrated in Figure 3.3(a) [71]. Liquid polyimide (PI-2574 from HD Microsystems) was spin-coated on a 50-nm thick plasma-enhanced chemical vapor deposited Si_3N_4 sacrificial layer on silicon. The 15μm-thick spin-coated polyimide (PI) affords a smooth surface with root-mean-square (RMS) roughness of < 1 nm. The PI coated film is soft-baked at 200 °C for 30 min and subsequently cured at 300 °C for 1 hr under a nitrogen atmosphere. An array of gate electrodes were patterned on a PI-coated silicon substrate by electron-beam lithography, evaporation, and liftoff. A high-k dielectric of 20-nm thick Al_2O_3 is deposited by ALD. A CVD grown graphene film from a Cu/SiO_2/Si growth substrate was then transferred via a conventional wet-transfer process using ammonia persulfate to etch the copper [69]. Oxygen plasma reactive-ion-etching (RIE) was used to pattern the active channel region while removing the superfluous graphene and ensuring channel isolation. Source and drain electrodes were defined to complete the device fabrication. The gate and source/drain metals are Ni/Au(10 nm /40 nm) and Ni(50 nm) respectively. Lastly, buffered oxide etchant (BOE 6 : 1) was used to strip the Si_3N_4 sacrificial layer and release the flexible polyimide film from the underlying silicon substrate. The second type of EGFET is using a cured PI film as the supporting flexible substrate. The 3D image of the fabricated 10-finger multi-finger EGFET unit-cell is illustrated in Figure 3.4(a). In brief, an array of gate electrodes were patterned directly on PI by electron-beam lithography (EBL), evaporation, and lift-off. A high-k dielectric of 15-nm thick Al_2O_3 was deposited by ALD with estimated gate-oxide capacitance of 405 nF/cm^2. Another EBL followed by wet-etching of Al_2O_3 by a 1 : 3 diluted solution of H_3PO_4 : DI water isolated each device while leavinglocal dielectric islands for the channel. High quality CVD monolayer graphene film was then transferred via the conventional poly (methyl methacrylate) (PMMA) wet-transfer process using ammonia persulfate to etch the supporting copper foil [68-69]. The active channel area of graphene was then patterned, and source and drain electrodes were defined to complete the device structure. The top of the sample was then covered with a 30-nm thick plasma-enhanced chemical vapor deposited (PECVD) Si_3N_4 layer to protect the multi-finger EGFETs from the outside environment. The channel length and width for each finger were fixed at 0.5μm and 20μm, respectively. Devices with both 10-finger and 18-finger configurations were prepared resulting in effective channel widths of 200μm and 360μm, respectively, on the same PI substrate.

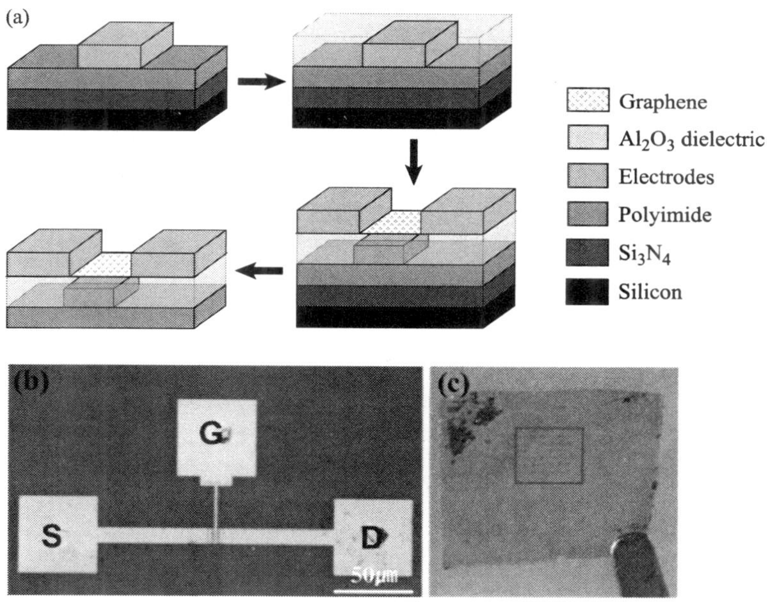

Figure 3.3. Embedded-gate graphene FET

(a) Illustration of the EGFET fabrication process on spin-coated PI on a Si_3N_4/Si substrate.
(b) Optical image of a completed EGFET. The device length and width are 4 and 8μm respectively.
(c) Photograph of the flexible substrate with an array of EGFETs. The highlighted square shows the array of devices.

The optical image given in Figure 3.4(b) shows the freestanding PI with an array of flexible MEGFETs bent with fingers. The inset shows a 3 × 3 array of the unit cells. Figure 3.4(c) is a high-resolution image of the active channel area obtained with an atomic force microscope (AFM) revealing a 10-finger electrode configuration.

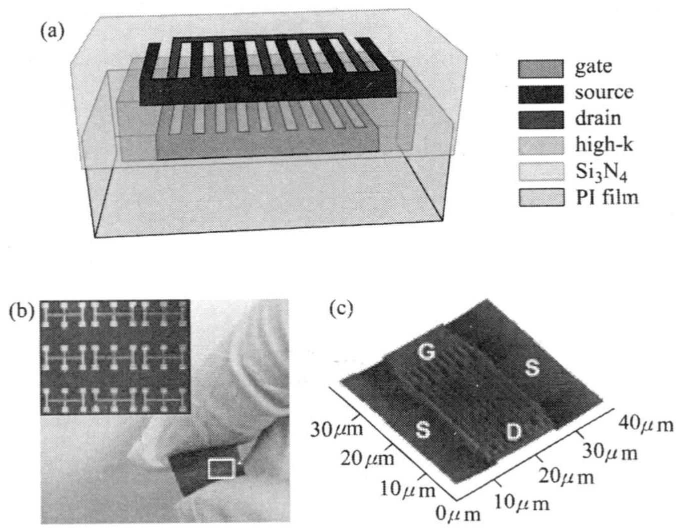

Figure 3.4. Multi-finger EGFET fabricated on plastic substrates.

(a) 3-D image of the complete device of a 10-finger unit cell.
(b) Optical image of the sample. The device area is highlighted by a white rectangle. Inset shows an array of the unit cells with gate (G), source (S), and drain (D) pads.
(c) AFM image of the active channel area revealing a device with 10-fingers

Modified Embedded-Gate Graphene Field-Effect Transistors

The fabrication process for the modified EGFET is as follows. h-BN (from Momentive Performance Materials) wasexfoliated on silicon wafers with 285-nm thick thermal oxide; the thickness of the h-BN flake was confirmed with AFM to be 19 nm. The sample was first annealed under nitrogen at 300 °C for 1 hour to achieve a smooth h-BN surface and remove residual adhesiveon its surface to define a better interface between gate electrodes and h-BN films. The simplified process flow is given in Figure 3.5(a). In brief, gate electrodes were patterned over annealed h-BN films on SiO_2/Si substrates by EBL, evaporation, and liftoff. Liquid-type PI (PI-2574 from HD Microsystems) was spin-coated on the patterned sample and cured under nitrogen to capture the device structures within PI. Thermal oxide was then etched by buffered-oxide-etchant (BOE) and the PI films with embedded gates and dielectrics were released (capture-release process). The released sample was then annealed again under nitrogen to provide a clean high-quality interface between graphene and dielectrics (two-step annealing process). Monolayer graphene film grown by chemical vapor deposition (CVD) process was transferred via the poly (methyl methacrylate) (PMMA) assisted wet-transfer process using ammonia persulfate as a copper etchant [69]. Redundant area of graphene film was removed by oxygen plasma etching. Finally, source and drain electrodes were defined to complete the device structure (Figure 3.5b). A high-resolution image over the active channel area and an optical image of the free-standing PI substrate with the device are shown in Figure 3.5(c) and 3.5(d), respectively. Figure 3.5(e) shows an AFM topographical scan of the h-BN film after anneal and Figure 3.5(f) compares the surface roughness of several gate dielectrics for GFETs on plastics. Previous works in our lab were focused on improving the surface quality of plastics by applying additional liquid-type layer and/or depositing high-k dielectrics over flexible films. The surface roughness has been remarkably improved down to 1~2 nm using these methods, however, its roughness is still higher than monolayer graphene, which can introduce mechanical strain and result in band structure deviation and transport degradation [70-73].

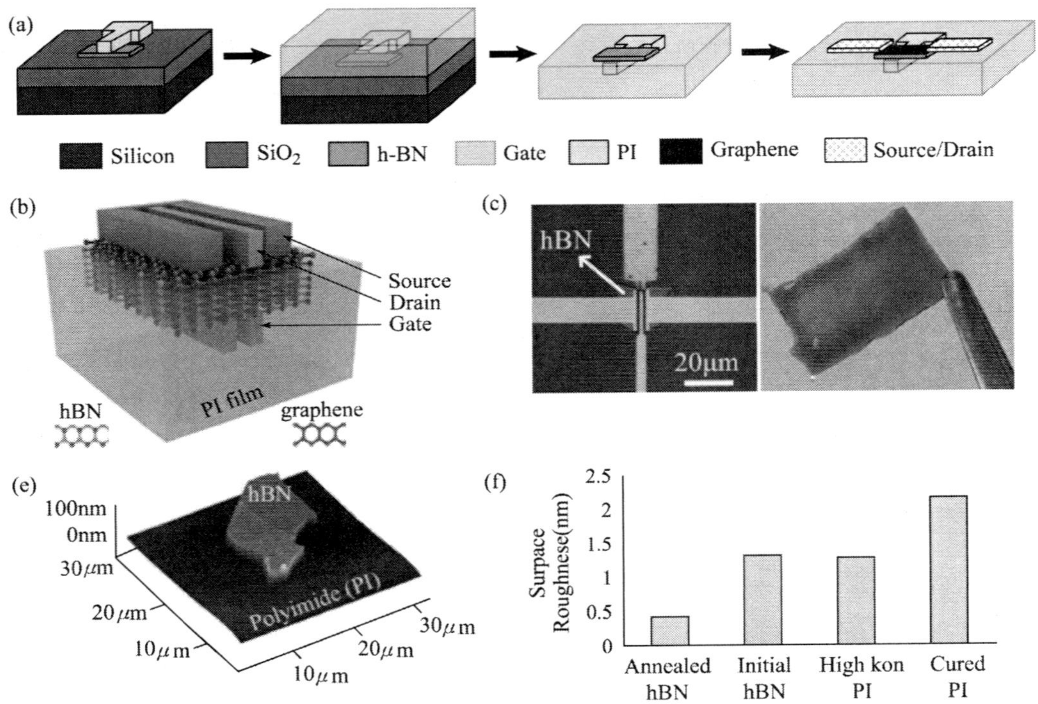

Figure 3.5. Illustration and images of fabricated EG-hBN GFET.

(a) Simplified process flow.
(b) Illustration of a fabricated device.
(c) Optical image of the channel area with a dual-finger embedded-gate GFET.
(d) GFETs on a PI film.
(e) Surface of the smooth h-BN on flexible PI as measured by AFM.
(f) Comparison of different surface roughnesses for graphene integration on flexible PI. The annealed h-BN surface features smoothness comparable to interlayer spacing in graphene.

Comparison of Device Structures

Three device structures, namely, top-gate GFET (TG-GFET), embedded-gate GFET with high-k dielectric (EG-highk / GFET), and embedded-gate GFET with h-BN dielectric (EG-hBN / GFET), which is the modified version of the previous EGFET to integrate h-BN into PI films, were fabricated on PI and evaluated in this work. Due to uneven surface topography of polymeric films, additional treatments are necessary to achieve the best lithographic resolution and registration [61]. In brief, PI films were spin-coated with liquid polyimide (PI2574 from HD Microsystems) and cured. TG-GFETs were prepared directly on these substrates, while others require more processing steps before graphene-transfer; ALD Al_2O_3 or exfoliated h-BN films were prepared as gate dielectrics. Optimized annealing was performed at 300 ℃ to improve the surface smoothness of h-BN films under nitrogen ambient. Figure 3.5(f) compares the roughness of the surfaces evaluated in this work. While the surface of as-exfoliated h-BN film is comparable to that of high-k on PIs, it is further improved by annealing as shown in the figure. After annealing, h-BN flakes on PI show the root mean square (RMS) roughness ~0.4 nm, comparable to the inter-layer spacing and indicative of an almost perfect residue-free interface. Table 3.1 compares the merits of the three device structures, investigated in this work. TG-GFET has an issue with the need for insulating seed layer prior to deposit ultra-thin high-quality dielectric over channel, which makes it difficult to scale down the oxide thickness. For embedded gate devices, oxide scaling is not a critical issue. While h-BN offers the best interface for GFET, substantial research has to be done to synthesize large-area uniform (monolayer to few layer) h-BN films. This leads to the conclusion that while embedded-gates with h-BN dielectric is ideal, EG-highk / GFET is the best practical device structure for now because further synthesis progress is needed for the integration of h-BN/graphene layers.

Table 3.1 Comparison of 3 flexible GFET device structures

Device structure / Deviceprocess	TG-GFET	EG-hBN /GFET	EG-high k/GFET	Comments
Finger scalability in a single layer	No	Yes	Yes	Realizing large # of fingers is straight-forward without the need for 2nd inter-connect level.
Gate dielectric seed layer	Needed	None	None	Seed layer is needed for even coating of high-k on graphene. Not needed for em-bedded-gate.
Gate dielectric scalability	No	Yes	Yes	For top-gate devices, seed layer prevents ultra-thin dielectrics (< 5 nm).
Automated gate dielectric process	Yes	No	Yes	Currently, growth and transfer of uniform large-area sheets of h-BN is not available.
Expected GFET performance	Good	Excellent	Very good	h-BN gate dielectric is expected to yield the best GFET performance

Chap 4

Design Considerations for Optimization

- Contact Resistance
- Embedded-Gate Electrodes
- Passivation Techniques for Embedded-Gate Structures
- ALD Dielectrics
- PECVD Dielectrics
- Integration of Organic Polymers on Graphene
- Thermal Management

The two most important metrics to quantify the high frequency response of the RF device are the transit frequency, f_T, and the maximum oscillation frequency, f_{MAX}. These are highly dependent on intrinsic and extrinsic device parameters. In order to achieve the highest RF performance, we need to take a look at which parameters affect the metrics. The following equations [(1) − (3)] are for (1) the intrinsic transit frequency, (2) the extrinsic transit frequency, and (3) the maximum oscillation frequency [74].

$$f_{T,int} = \frac{g_m}{(2\pi C_{gs})} \tag{1}$$

$$f_{T,ext} = \frac{g_m}{2\pi r(C_{gs} + C_{p,gs} + C_{p,gd})[(R_{p,s} + R_{p,d})g_{d+1}] + C_{p,gd}\, g_m (R_{p,s} + R_{p,d})} \tag{2}$$

$$f_{MAX} = \frac{f_T}{\sqrt{2g_d(R_{p,s} + R_{gate}) + 2\pi f_T C_{p,gd} R_{gate}}} \tag{3}$$

where g_m is the transconductance, g_d is the drain-conductance, C_{gs} is the gate capacitance, $C_{p,gd}$ and $C_{p,gs}$ are the parasitic gate-to-drain and gate-to-source capacitances, $R_{p,s}$ and $R_{p,d}$ are the parasitic resistances at the source and drain electrodes, and R_{gate} is the gate resistance. The reduction in the channel length is the most straightforward way to improve the intrinsic f_T since the intrinsic f_T scales proportional to $\frac{1}{L}$ [75]. The extrinsic f_T can be dominated by the device parasitics. Currently, the source / drain electrodes for our devices show much higher than the values frequently obtained from other reports on GFETs; the contact resistances from our devices range from 1 k $\Omega \cdot \mu\text{m}$ to 5 k $\Omega \cdot \mu\text{m}$, while other reports show mostly less than 1 k $\Omega \cdot \mu\text{m}$ even less than 100 $\Omega \cdot \mu\text{m}$ This requires more investigations on the interface between graphene and metal electrodes and proper designs to reduce the contact resistance. The power gain, f_{MAX}, is also the importantmetric for RF devices. The improvements on f_{MAX} can be achieved by (1) reduction in parasitics by better layouts or fabrication, and (2) reduction in the gate resistance.

Contact Resistance

The contact resistance of graphene transistors consists of the following components: the metal electrode interconnection, the un-gated graphene resistance, and the resistance at the metal-graphene interface. By utilizing the highly conductive metals like gold as the electrodes, the first term can be improved, while the second term can be neglected with the proposed gate-overlap device structure. This makes the last term, the resistance at the metal-graphene interface, becomesthe most critical component determining the overall contact resistance of the fabricated device. The low quality interface, leading to the high contact resistance, is the fundamental limiting factor on the electrical performance of graphene devices. It can degrade the transconductance and the cutoff frequency of GFETs by more than an order of magnitude [76]. Previously, the fundamental charge transport physics on the graphene metal contacts and interface quality were investigated to provide better understanding on the graphene-metal contact resistance [77-81]. The chemical residues can be introduced onto graphene films during the transfer of graphene films from Cu foils to desired substrates for fabrication. PMMA used during this transfer process can reside on the graphene surface and it is difficult to be completely removed from the surface by acetone during the cleaning process. The detrimental effects from these residues on graphene devices result in poor electrical performances with low field-effect mobilities. High vacuum annealing (1×10^{-9} mbar) at 300 °C for 3 hours has been suggested to remove them [77]. However, annealing samples at 300 °C for extended time is not feasible for flexible substrates which are not thermally stable. Another approach suggested using an Al sacrificial layer prior to lithographic steps defining contact patterns [78]. Inthis case, the effect from the chemical residue can be minimized since the chemical is not touching the graphene surface at any time during the entire lithographic procedures, which results in up to 10X improvement in a contact resistance. The selection of contact metals also plays a role determining the electrical performances of graphene FETs. A theoretical study using density functional theory suggested that the doping of graphene can be induced by adsorption on metal films and the cross-over of doping from p-type to n-type can be also expected by utilizing different metals (more importantly utilizing their work-functions) [79]. Another experimental study also showed the p-n junction formation at the interface of graphene/metal can be induced by utilizing different metals; high work-function metals afford a more transparent interface (p-type to p-type) for hole transport in contrast to electron transport that involves a p-n junction interface with an additional contribution to the contact resistance [82]. This effect can lead to the asymmetry in the contact resistance between electron and hole transports [83-84]. Recently, another publication suggested an interesting idea to further reduce those values. In order to improve the carrier injection at the metal-graphene interface, the contact regions of graphene devices are patterned into multiple cuts [85]. By simply modifying the design under the contact areas, two carrier injecting mechanisms, "edge-contacted" injection and "top-contacted" injection, contribute to lower the contact resistance values by 22～32% compared to those values without patterns [85]. In our work, the contact resistances coming from source and drain metal electrodes interfacing with the graphene channel offers a great portion of the total resistance, limiting its extrinsic electrical performance. In order to access the complete usefulness of the fascinating properties of graphene, we introduced the patterned contacts to improve the carrier injection at the interface between metal contacts and the graphene channel.

The total contact resistance is the parallel combination of the interface resistance at the metal-graphene interface, R_{Metal}, and the resistance coming from the edge-injection effects, R_{Edge}.

$$R_{Total} = R_{Metal} // R_{Edge} \qquad (4)$$

The equation given above can be also expressed in terms of the conductance for analysis.

$$\rho_{Total} = \rho_{Metal_{Total}} + \rho_{Edge_{Total}} \tag{5}$$

$$\rho_{Total} = \rho_{Metal/width} \cdot W_{Channel} + \rho_{Edge/width} \cdot W_{Edge} \tag{6}$$

where ρ_{Total} is the total conductance through the contacts, $\rho_{Metal_{Total}}$ is the contribution through the metal-graphene interface, and $\rho_{Edge_{Total}}$ is the contribution from the edge charge injection. Again, the equation can be further expanded considering the channel width, $W_{Channel}$, and the electrode width for the edge injection, W_{Edge}, and the per-width partial conductance from the metal-graphene interface and the edge-injection effect. In the experiment, we fabricated devices with different contact patterns as given in Figure 4.1. Figure 4.1(a) is the conventional contact layout without edge-injection patterns. The device has a multi-finger (6 fingers) configuration with narrow metal electrodes for minimum edge-injection effect. Figure 4.1(b) is the proposed contact layout with line-type edge-injection patterns and Figure 4.1(c) is another proposed contact layout with array-type edge-injection patterns. The contributions from the metal-graphene interface and the edge-injection can be extracted from electrical measurements. Firstly, the devices are measured to extract the contact resistance values for each patterned structure. Then, a simple matrix calculation can be used to extract the individual contributions as follows. R_{Total} is extracted directly from the electrical measurements, while other parameters of $W_{Channel}$ and W_{Edge} are from the physical layouts.

$$\rho_{Total} = \frac{1}{R_{Total}} \tag{7}$$

$$\begin{pmatrix} W_{channel_1} & W_{Edge_1} \\ W_{channel_2} & W_{Edge_3} \end{pmatrix} \cdot \begin{pmatrix} \rho_{Metal/width} \\ \rho_{Edge/width} \end{pmatrix} = \begin{pmatrix} \rho_{Total_1} \\ \rho_{Total_2} \end{pmatrix} \tag{8}$$

$$\begin{pmatrix} \rho_{Metal/width} \\ \rho_{Edge/width} \end{pmatrix} = \begin{pmatrix} W_{channel_1} & W_{Edge_1} \\ W_{channel_2} & W_{Edge_3} \end{pmatrix}^{-1} \cdot \begin{pmatrix} \rho_{Total_1} \\ \rho_{Total_2} \end{pmatrix} \tag{9}$$

From the electrical measurements, the total contact resistances are extracted as follows.

$$R_{Total,\,Type-I} = 93.03\,\Omega$$

$$R_{Total,\,Type-II} = 53.67\,\Omega$$

$$R_{Total,\,Type-III} = 40.28\,\Omega$$

After calculating the partial contributions, the resistances from each component are also extracted.

$$R_{Metal} = \frac{1}{\rho_{Metal/widht} \cdot W_{Channel}} = 108\,\Omega$$

$$R_{Edge,\,Type-II} = \frac{1}{\rho_{Edge,\,Type-II/width} \cdot W_{Edge}} = 114\,\Omega$$

$$R_{Edge,\,Type-III} = \frac{1}{\rho_{Edge,\,Type-III/width} \cdot W_{Edge}} = 63.9\,\Omega$$

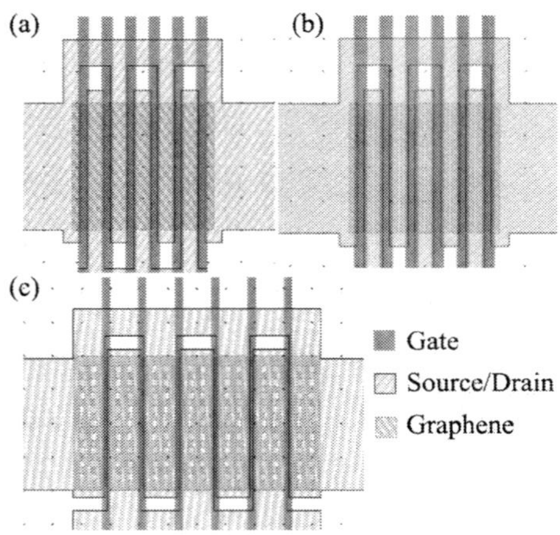

Figure 4.1. Contact electrodes patterns.

(a) Type-I. Conventional contacts without edge-injection patterns.
(b) Type-II. Contacts with line-type edge-injection patterns.
(c) Type-III. Contacts with array-type edge-injection patterns.

The patterned contacts, Type-II and Type-III, offered lower contactresistances compared to the conventional one, Type-I. The data confirm the array-type (Type-III) patterned contacts offer the effective reduction in the total contact resistance while providing an additional 25% reduction in the contact resistance compared to the line-type (Type-II) patterned contacts.

Embedded-Gate Electrodes

Since the power gain, f_{MAX}, strongly depends on the gate resistance, we investigated different options for preparing the embedded-gate electrodes. While selecting the materials, several things to be considered in this process are i) surface oxidation of the metal, which leads to higher resistance over time, ii) high electrical conductivity for lower resistance, iii) structure of the embedded-gates in order to minimize the device parasitics, and iv) surface roughness over the metal film. The last requirement is crucial since the graphene devices are currently prepared on top of the gate metals covered with dielectric films to have an embedded-gate structure to access high electrical performance as mentioned in the previous chapter. In this work, we compared Ti/Pd (2 nm/80 nm), Ti/Ag/Pd (2 nm/70 nm/10 nm), Ti/Pd/Au (2 nm/10 nm/70 nm) for preparing gate electrodes. The first one stands for the reference device that we routinely utilize to implement graphene field-effect transistors. Palladium offers reasonablylow resistance while providing smooth surface where short-channel devices are easily implemented. Silver and Gold are considered in order to further reduce the gate resistances. Since the surface oxidation of Silver is detrimental while that has negligible effects on Gold, we covered the Silver electrodes with an additional Palladium of 10 nm to prevent this surface oxidation. The electrical conductivities of the three candidates are compared via the TLM (transmission line method) measurements to extract their sheet resistance values. The surface roughness values are also compared using an atomic force microscopy scanning over 15μm × 15μm area for each candidate. Figure 4.2 shows the comparison of these candidates.

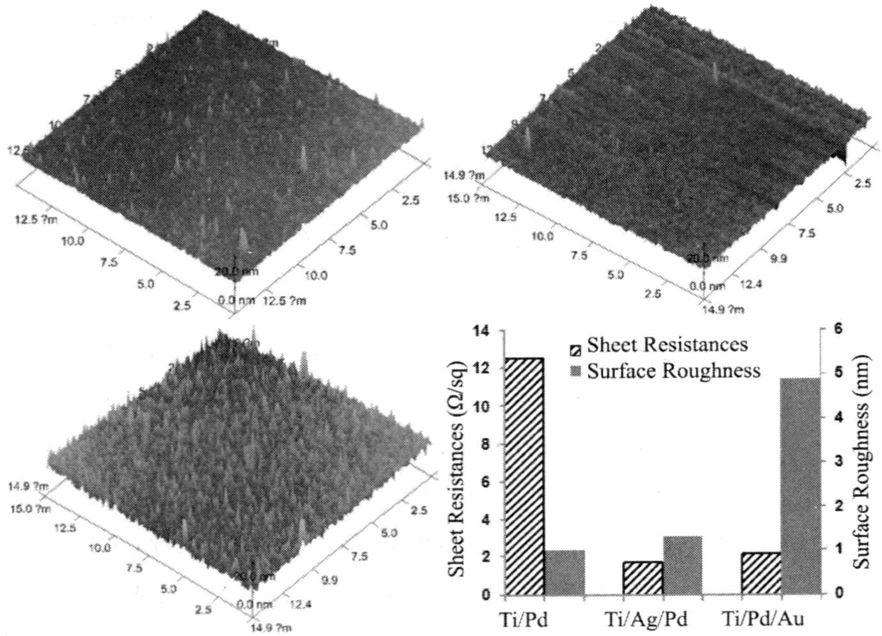

Figure 4.2. Comparison of gate electrodes.

(a) AFM image of Ti/Pd (2 nm / 80 nm).
(b) AFM image of Ti/Ag/Pd (2 nm / 70 nm / 10 nm).
(c) AFM image of Ti/Pd/Au (2 nm / 10 nm / 70 nm). All images are scanned over 15μm × 15μm.
(d) Comparison of sheet resistances and surface roughness values for three candidates.

As shown in Figure 4.2(d), Ti/Ag/Pd offers the highest electrical conductivity along with a reasonable surface roughness making it as the most fascinating choice among the compared candidates. Ti/Pd offers the best (lowest) surface roughness while providing relatively higher sheet resistance, which may severely limit the power gain performance of the fabricated device. The last option of Ti/Pd/Au also offers a reasonable sheet resistance, however the un-even surfacelimits its usefulness as the embedded gate electrodes. Another way to further improve the electrical conductivity of the gate electrodes is increasing the thickness of the metal stacks. The direct deposition of high metal stacks leads to proportionally lower resistance at the cost of higher parasitic capacitances as shown in Figure 4.3(a). The better way to deal with this issue would be introducing the trench structure where the gates electrodes are fully embedded (Figure 4.3b). The extension length of source and drain electrodes over the gate electrodes is 200nm for the prepared device leading to the parasitic overlap capacitance, $C_{OV,TOP}$. With 50 nm-thick gate electrodes, the parasitic component of the side-wall overlap capacitance, $C_{OV,SIDE}$, increases the overall parasitic capacitance by additional 25%. Increase in the thickness of this metal stack would also increase this parasitic part leading to the limited

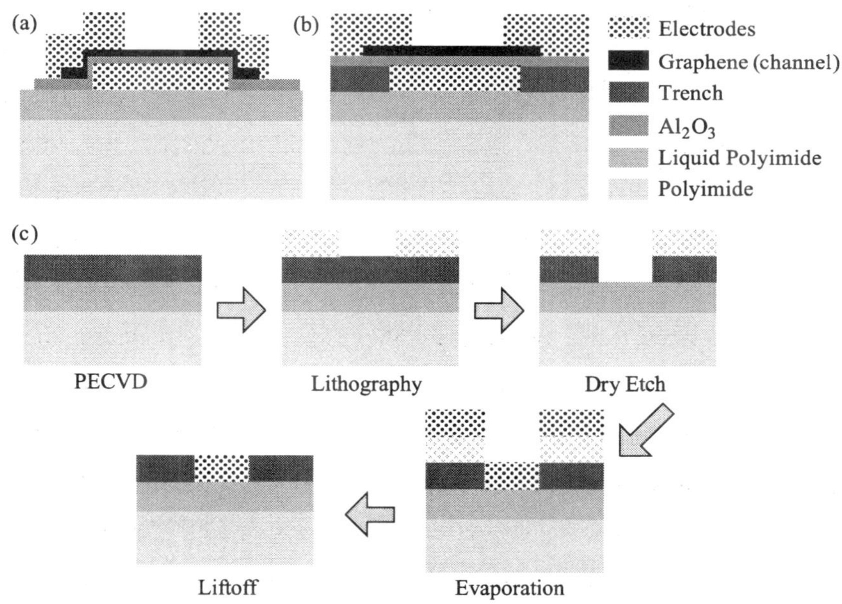

Figure 4.3. The embedded gate structures.

(a) Direct deposition of gate electrodes on the substrate.
(b) Gate electrodes embedded in dielectric trench structures.
(c) Simplified process steps to prepare the self-aligned gate-electrodes embedded in the dielectric trench.

extrinsic high frequency performance. In order to avoid this trade-off, we suggest the trench-type gate electrodes as shown in Figure 4.3(b) to remove the side-wall overlap capacitance while lowering the gate-overlap capacitance and also effectively increasing the cutoff frequency performance of the device. Figure 4.3(c) shows the simple self-aligned patterning technique to define the dielectric trench and deposit the embedded gates in one lithography step. The electron-beam lithography is used to define the trench patterns and the fluorine-based dry etch etches the nitride trenches where the metal electrodes are refilled via the electron-beam evaporation and the subsequent liftoff process.

The successful patterning of trenches strongly depends on the high-aspect ratio reactive-ion-etching and the high-selectivity between the e-beam resist (PMMA) and the dielectric material used for the trench structures. Here, we used the plasma-enhanced chemical vapor deposited silicon nitride for the dielectric material. This process enables the deposition of thick dielectric films and its low-temperature process (250 °C) is compatible with the flexible substrates. For the reactive ion etching of nitrides, CF_4 is commonly used, however it also shows a very high etch rate for PMMA. Introducing Ar can significantly improve the etching selectivity while lowering the etch rate for PMMA. We used the 790 Plasma Therm RIE Etcher during this process. The process condition is as follows. 50 sccm of CF_4, 20 sccm of Ar, RF power of 100 W, and the pressure of 80 mTorr. Figure 4.4 shows the results. Figure 4.4(a) is the AFM image of the patterned trenches showing the area for the 6-finger gate electrodes and the pad. Figure 4.4(b) is another AFM image of the same area after refilling metals using the electron-beam evaporation of Ti/Ag/Pd. The height of the metal stack and the depth of the trench are determined to be the same (80 nm) to provide the flat surface where the devices are to be implemented. The etch rate for the nitride and PMMA are 114 nm/min and 88 nm/min, respectively, offering the good etch selectivity of over 1.29 : 1. The thickness of the resist layer is 250 nm.

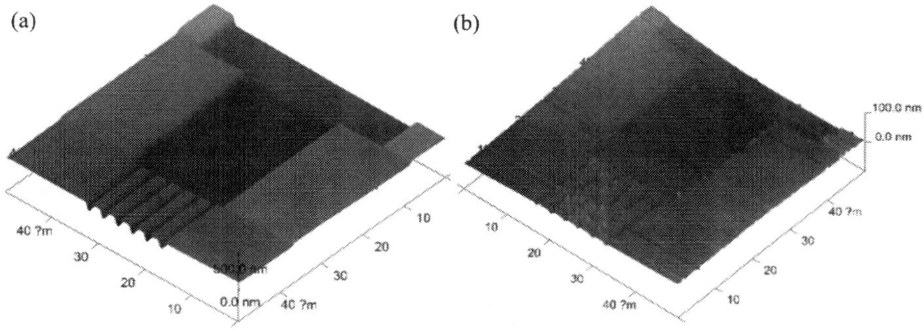

Figure 4.4. The trench refilling process.

(a) AFM image of the patterned trenches showing the area for the 6-finger gate electrodes and the pad.
(b) AFM image of the same area after refilling metals using the electron-beam evaporation of Ti/Ag/Pd.

Passivation Techniques for Embedded-Gate Structures

While the embedded-gate GFETs outperform the conventional top-gate GFETs as described in Table 3.1, the still remaining issue is the passivation of exposed graphene channel. The proper passivation is required for 1) improved hysteresis, 2) preventing degradation during device operation, 3) mechanical robustness, and 4) protection overdirect liquid exposures. Several approaches have been reported for GFETs on conventional rigid substrates: bilayer (organic/inorganic) passivation, PECVD silicon nitride directly on graphene, and hydrophobic organic polymer spincoated on graphene [86-88]. Here, we have investigated different options: ALD high-k dielectrics, conventional PECVD Si_3N_4, N_2-rich PECVD Si_3N_4, bilayer films of N_2-rich PECVD Si_3N_4 covered with CYTOP fluoropolymer.

ALD Dielectrics

The first one we investigatedis using ALD dielectrics. ALD films offer high quality dielectric films good for both gate oxides and device passivation layers. Several high-k dielectric materials are available including Al_2O_3, HfO_2, and ZrO_2. However, this process requires the deposition of an insulating seed layer prior to deposit dielectrics over graphene due to the chemical inertness of graphene surface and not reliable [89]. While preparing this film as the gate dielectric, we can still avoid this issue by adopting the embedded-gate FET structure [84, 90].

PECVD Dielectrics

PECVD is another option and oxides and nitrides are routinely deposited on substrates as interlayer dielectrics, passivation layers, or gate dielectrics. PECVD oxideis excluded among the feasible candidates for the device passivation since it requires the deposition involving oxygen radicals, which effectively etch the graphene channel leaving pin-holes after the deposition [87]. The conventional PECVD recipe for Si_3N_4 is also evaluated. However, still the graphene channel has been degraded during plasma process significantly degrading the field-effect mobility of the device after passivation. The modified recipe using N_2-rich PECVD could prevent this degradation in its electrical properties and also offers good mechanical robustness. When the nitride-passivated sample is exposed to liquids, the nitride passivation layer fails to protect the underlying FET device and the device experienced significant reduction in quality over time, which will be covered shortly in the later chapter.

Integration of Organic Polymers on Graphene

Several publications suggest the usefulness of organic polymers introduced on graphene surfaces as promising options to enhance or modify the electrical properties of graphene devices prepared on rigid substrates [91-94]. Self-assembled monolayers (SAMs) are molecular thin films formed via chemical or physical interactions of molecules with a substrate without the utilization of high-temperature or vacuum processes [94]. B. Lee et al reported that they could successfully grow uniform and stable fluoroalkyl silane self-assembled monolayer (FTS SAM) on the graphene surface, which leads to a high level of hole doping effect with a density of up to $1.5 \times 10^{13}\ cm^{-2}$ [91]. Even though the field-effect mobility from the graphene device degraded significantly, this approach highlights the potential of SAM layers for graphene electronics with the capability of modifying the electronic properties of devices. Another report focused on the effect of the hydrophobicity of supporting layers; a thin, hydrophobic self-assembled layer of hexamethyl disilazane (HMDS) was introduced on the supporting substrate prior to the fabrication of graphene devices [92]. From the results, the hysteresis coming from the adsorption of dipolar substances on graphene surface could be minimized or improved via the HMDS treatment on substrates. This suggests that SAMs are also useful to access the superior electronic properties of graphene with low intrinsic doping level, which is largely dependent on processing conditions and ambient conditions, in a reproducible manner.

More recently, we reported spin-on-fluoropolymers offer a simple and fascinating method to achieve a restorative effect on electrical properties of graphene field-effect transistors [93]. The strong polar nature of the C-F chemical bond in the spin-on fluoropolymer deposited on graphene devices results in the favorable improvements in all aspects of electrical properties of graphene devices including high field-effect mobility, low Dirac voltage close to zero, and very low residual carrier density. While aforementioned approaches may offer promising options to change or enhance the electrical properties of graphene, low boiling points and melting points of most organic polymers prevent the direct introduction of these materials on graphene FETs prepared on plastic substrates where the temperature inside the active channel area can become significantly higher than those values. Here, as an alternative way, we utilize a bi-layer structure of N_2-rich PECVD Si_3N_4 and hydrophobic fluoropolymer. This preserves the electrical performance of the graphene devices along with the additional benefit of protection over liquids and its results on the electrical performance will be covered in the later chapter.

Thermal Management

Due to poor thermal stability of most flexible substrates, the thermal management for these films has crucial effects on realizing reliable high performance flexible devices. Especially, the 2-D material operating under high current density like graphene, this becomes more significant issue. Figure 4.5 exemplifies this issue on flexible GFETs. The physical deformation of the device (shown in Figure 4.5b) results from the fact that the GFET is implemented on the thermally unstable material with low thermal conductivity (polyimide). Due to both low glass transition temperature of plastic substrates and elevated channel temperature at high bias, flexible films tend to melt or deform under these conditions. The current saturation, which is essential to obtain high f_{MAX}, can be achieved under high lateral field and, partly for this reason, current-saturating GFETs on plastics are not easily realized while similar devices are reported frequently on rigid substrates. This needs some ways to prevent direct heat transfer from the channel to the substrate. Diverse solutions on this matter are currently under active investigation: direct heat spreading via electrodes, high thermally conductive dielectric heat spreading layers, and dielectric heat insulator to protect the substrates.

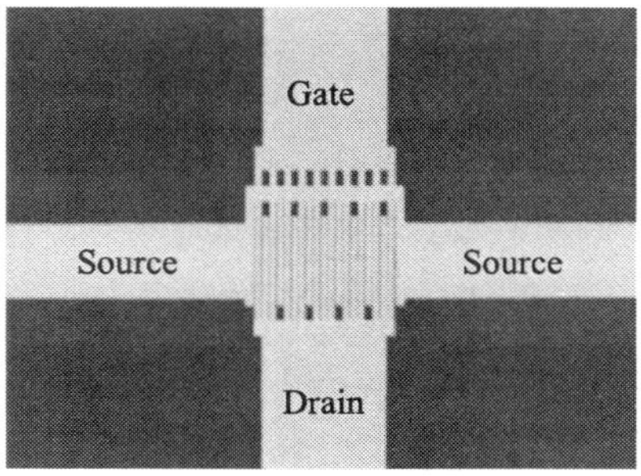

(a)

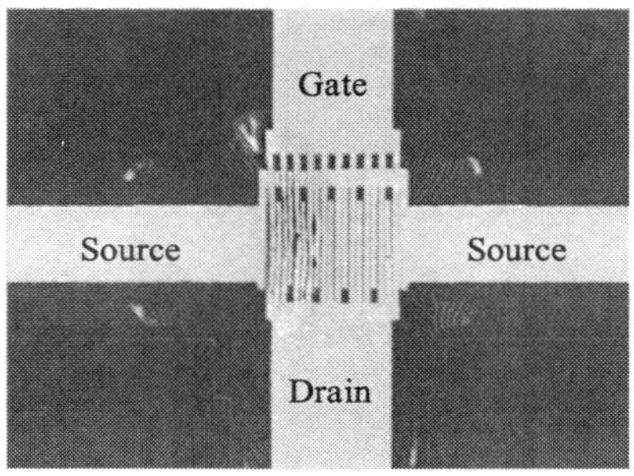

(b)

Figure 4.5. Thermal effects on GFETs under high lateral electric fields.

(a) SEM image of the GFET before applying high electric fields. The devicehas a multi-finger electrodes configuration and the graphene channel is located at the center of the image.

(b) SEM image of the same device (damaged) after applying high electric fields.

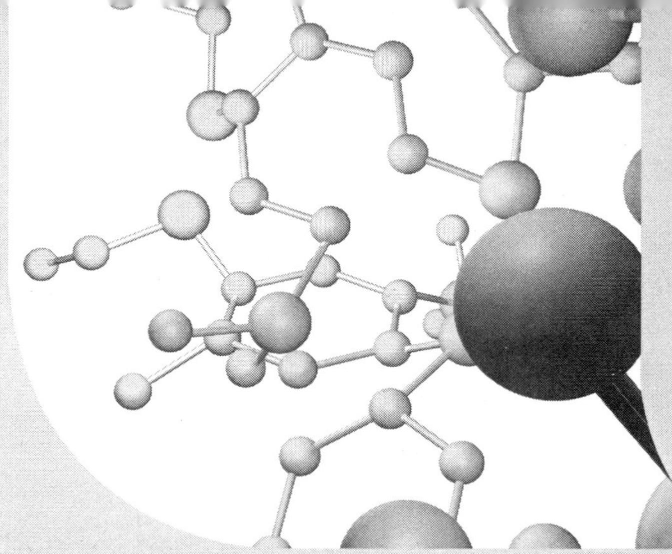

Chap 5

Experimental Results : Electrical / Mechanical Performances

- DC Characteristics of Graphene Field-Effect Transistors
- High Frequency Response (f_T and f_{MAX})
- Mechanical Bendability
- Liquid Exposure

DC Characteristics of Graphene Field-Effect Transistors

Electrostatic measurements for hBN-EGFET are presented in Figure 5.1. The device parameters were extracted and showed good agreements with a widely used diffusive transport model [89]. The extracted mobilities were 2,324 $cm^2/V \cdot s$ and 2,307 $cm^2/V \cdot s$ for holes and electrons, respectively. While the quality of CVD graphene on evaporated Cu film is very good, however, the graphene domains are smaller than those achieved on Cu foil. Hence for our devices, transport is likely occurring over several grain boundaries which limits the carrier mobility. Low impurity concentration ($< 5 \times 10^{11}$ $/cm^2$) owing to high quality dielectric interface, allows symmetric charge transport and high mobility. High work-function metal electrodes (50-nm Au) form excessive junction resistance for electron transport and reduce the drive current. Relatively high extracted contact resistance of 800 $\Omega \cdot \mu m$ indicates that the current density can be further enhanced by improving the metal-channel interface. Current saturation, which also offers record-high current density from flexible GFETs, was observed for the first time on flexible plastics as shown in Figure 5.1(b) and 5.1(c) with strong agreement with a validated GFET circuit model data reported on previous works on flexible substrates so far only provide the linear I-V relation without any current saturation [95-97]. We note that our device model includes both velocity saturation ($\sim 2 \times 10^7$ cm/s) and electron-hole charge crossover phenomena, which account for the current saturation in graphene [98]. The current saturation observed in our device is due to velocity saturation as determined from device modeling. h-BN offers high quality interface with low impurity density and high thermal conductivity; beneficial properties that enables access to the intrinsic graphene device behavior under high fields. This, in addition to low contact resistance from clean metal deposition is crucial for observing velocity and current saturation [98].

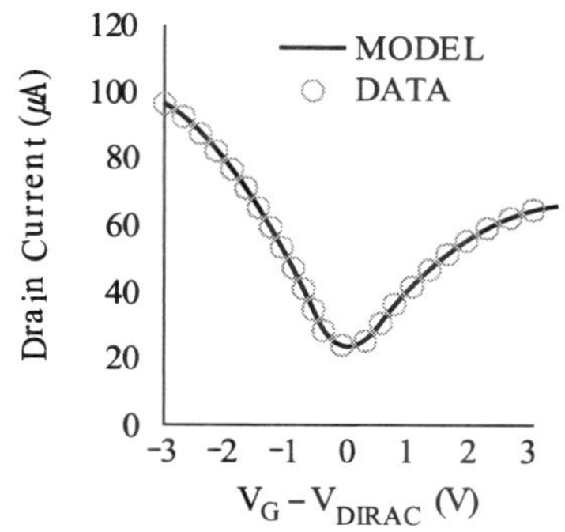

(a)

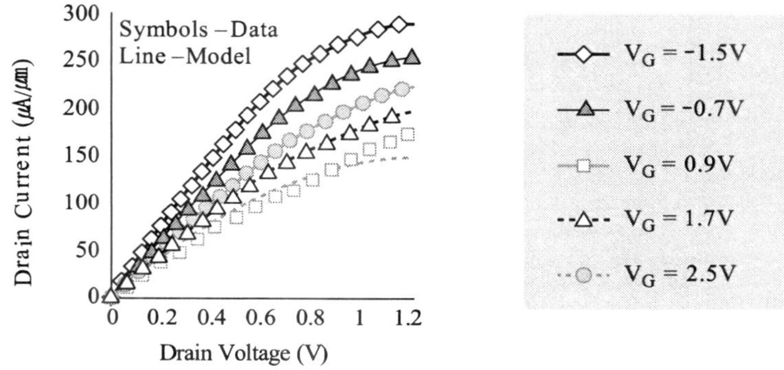

(b)

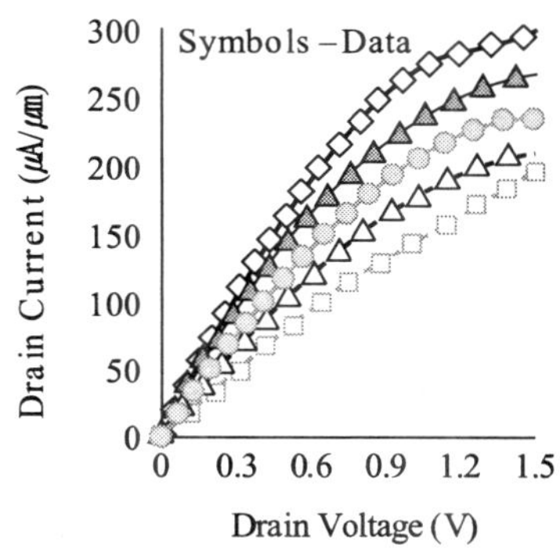

(c)

Figure 5.1. DC measurement of hBN-EGFET.

(a) I_D-V_G profile measured at V_D of 10 mV.

(b) I_D-V_D profile showing the current saturation.

(c) I_D-V_D profile at V_D up to high-field of 1.5 V.

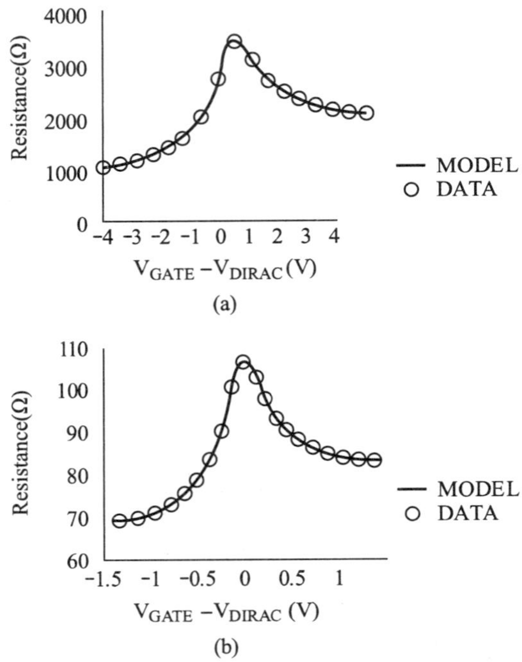

Figure 5.2. Electrical measurements.

(a) Gate modulation from the lithography-free GFET with the carrier mobility of 6600 $cm^2/V \cdot s$ for hole transport and 8000 $cm^2/V \cdot s$ for electron transport.

(b) Gate modulation of the EBL GFET showing the carrier mobility of 2800 $cm^2/V \cdot s$ for hole transport and 3900 $cm^2/V \cdot s$ for electron transport, respectively. The device has a channel length of 0.5μm, and effective width of 100μm.

Figure 5.2 shows the electrical properties of the multi-finger EGFETs evaluated under ambient conditions; the device is utilizing a modified CVD-process grown graphene films to achieve the highest electrical performance, which offers exfoliated quality large-area graphene films [99]. The gate modulation over the total resistance including contributions from the graphene channel and the metal contacts from the lithography-free device and the EBL patterned device are shown in Figure 5.2(a) and 5.2(b), respectively. The lithography-free devices have the same process flow (same metal electrodes and dielectrics) as the EBL patterned device with the exception of a global bottom-gate instead of patterned finger-type electrodes used in the other devices in order to identify the highest performance achievable with pristine, un-damaged (duringthe lithographic process introducing several chemicals and mechanical handling) graphene films. Data have good agreement with a low-field diffusive transport model for graphene transistors. The peak carrier mobilities of 8000 $cm^2/V \cdot s$ from the lithography-free device and 3900 $cm^2/V \cdot s$ from the EBL device are measured and confirmed with the device model. The asymmetry in the carrier transport comes from the use of high work-function metal as electrodes, which favors hole transport while introducing additionaljunction resistances for electron transport. The degradation in the carrier mobility from the latter device is coming from the undesirable resist residue left on the graphene channel, which requires further studies on an efficient resist removal and surface cleaning process compatible with flexible substrates. The hBN-EGFET, which offered the high symmetry in both electron and hole transports along with low impurity density, was also evaluated as a frequency doubler. The graphene frequency doubler has beenpreviously reported to offer intrinsically high spectral purity (> 90%) without filtering [95, 100]. The input and output terminals were terminated with a 50 Ω cable/load typical of RF circuits and the GFET was biased at the Dirac point. Figure 5.3(a) shows the graphene frequency doubler circuit. Figure 5.3(b) shows the output power at the doubled frequency as a function of the input power at the fundamental frequency. The solid line in Figure 5.3(b) is a visual guide with a slope of 20 dB/decade that indicates theexpected behavior of an ideal square law doubler. The inset is a photograph of the signal analyzer display. The conversion gain (CG) is −29.5 dB with an output power of −22.2 dBm, which was afforded by the low impurity interface betwyeen graphene/h-BN and the low inter-plane thermal conductivity of h-BN, which protected the underlying PI from thermally induced deformation and damage at high fields. This is the highest reported CG for flexible GFET and has nearly twice the output power compared to a recent graphene doubler from state-of-the-art graphene wafer technology. At higher input power, the even/odd symmetry from the transistor output characteristics introduces some non-linearity which degrades the doubler conversion loss and spectral purity [101].

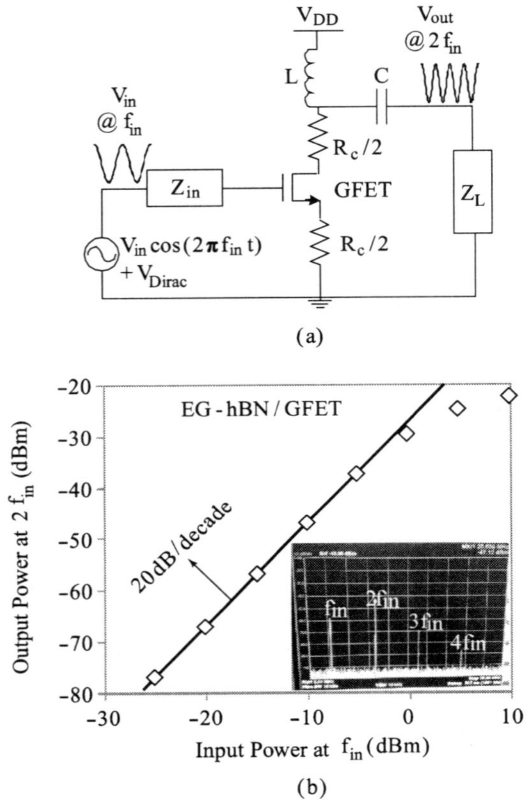

Figure 5.3. Experimental GFET frequency doubler circuit on flexible PI.

(a) Schematic for doubler evaluation.
(b) GFET output-input doubler characteristics (f_{IN}=10 MHz, V_D=0.5 V).

Figure 5.4 is the simulated performance assessment of a GFET frequency doubler with ultra-scaled gate dielectric consisting of a monolayer of h-BN. Figure 5.4(a) shows the impact of charge impurity density on the I_D-V_G curvature and gate modulation, parameters that are critical for doubler operation. Figure 5.4(b) is the dependence of the maximum CG on device parameters. In order to reach the theoretical limit of GFET doublers, several requirements have to be satisfied including high gate capacitance, low charge impurity density, and low contact resistance. A theoretical CG of ~ −4.5dB is predicted in 50 Ω systems indicating that optimized GFETs with scaled h-BN dielectrics can outperform ideal varistor diode doublers. The inset shows the dependence of the maximum CG on contact resistance. There is negligible impact for R_c < 100 Ω-μm.

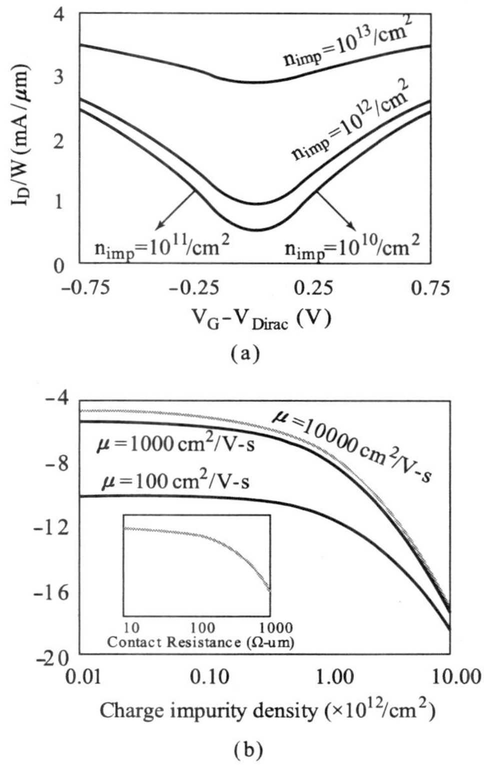

Figure 5.4. Simulated ultimate performance prediction of GFET frequency doubler with ultra-scaled gate dielectric. (monolayer h-BN, L=0.2μm and W=100μm. μ=10,000cm²/V-s)

(a) Detrimental impact of charge impurity density on gate modulation.
(b) The dependence of the max CG on device parameters. The inset is the dependence on contact resistance.

High Frequency Response (f_T and f_{MAX})

A central motivation for exploring graphene is its high-speed charge transport which can enable RF and high-frequency electronic applications on flexible substrates. In this light, we experimentally evaluated its scattering parameters (S-parameters) to determine the current gain (h_{21}) and cut-off or transit frequency (f_T) which is a key metric of device speed [102-104]. The as-measured S-parameters typically include delay contributions from parasitic capacitance and interconnect resistance that prevents direct evaluation of the intrinsic speed of charge carriers [104]. The parasitic components arise from the metal electrodes used in the RF device structure. To overcome this limitation, we employ a two-step de-embedding method that involves measurement of so-called *open* and *short* test structures identical to the device structure and fabricated on the same substrate but without the graphene active layer in order to subtract out the parasitic capacitance and resistance respectively [103-104]. The intrinsic current gain is shown in Figure 5.5(a) revealing an $f_T \sim 25$ GHz.

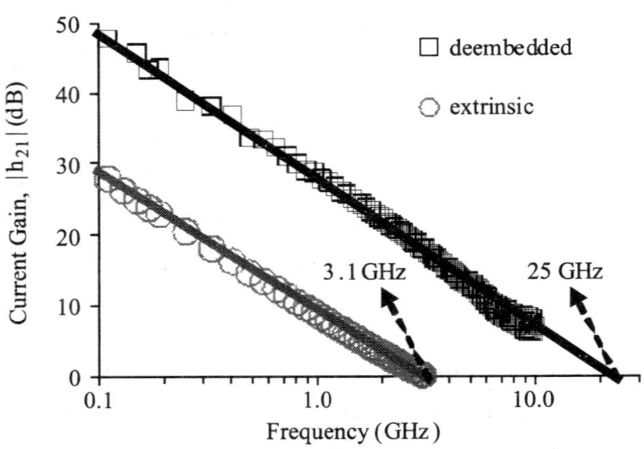

Figure 5.5. The extrinsic and intrinsic cut-off frequency measured from the device. The de-embedded current gain reveals an intrinsic f_T of 25 GHz for flexible GFET.

We note that the graphene contact resistance which represents a heterogeneous interface resistance is not removed by the de-embedding method. Hence, the extracted 25 GHz f_T reflects a conservative estimate of intrinsic speed especially in light of the relatively high contact resistance (3 kΩ-μm) in the present device. The extracted flexible GFET f_T is about 50% lower than the highest reported GFETs of similar channel length on conventional hard substrates, albeit the drain voltage of our flexible GFET is less than half of that on the hard substrate. The low thermal conductivity and glass transition temperature of plastic sheets are a barrier that restricts the fields and current densities necessary to probe the ultimate cut-off frequency on flexible substrates. Basic exploration of heat management materials (such as hexagonal boron nitride and other suitable thin films) is a matter of critical importance for advanced flexible electronics and warrants awareness and further studies. An f_{max} ~2.1 GHz was extracted from Mason's unilateral gain as given in Figure 5.6. The relatively low f_{max} is due to the well-recognized impact of un-optimized gate metal resistance and gate-drain coupling capacitance [104-105]. Applied research utilizing self-aligned T-gate structures on flexible substrates is needed in order to achieve substantial improvement in f_{max} [106]. The intrinsic cut-off frequency can also be estimated from the well-established analytical expression for field-effect transistors.

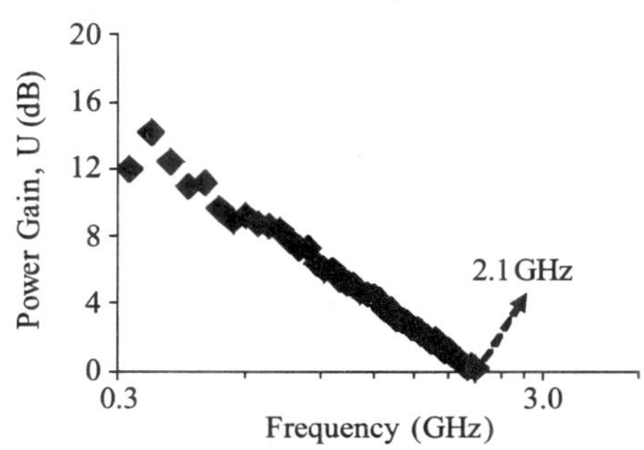

Figure 5.6. The power-gain frequency measured from the device.

$$f_t = \frac{g_m}{2\pi C_{gc}} \tag{10}$$

where g_m is the intrinsic transconductance and C_{gc} is the gate to channel capacitance. Normally, C_{gc} is taken to be the gate oxide capacitance (the parallel plate capacitance between the gate metal and the channel layer). However, owing to the relatively thin (10 nm) high-k dielectric used in our device, the contribution from the quantum capacitance (C_q) cannot be neglected. C_q can be considered approximately 2000 nF/cm^2 for the typical range of carrier densities ($1-10\times10^{12}$ cm^{-2}). This value was also previously used by Petrone et al [107]. The effective C_{gc} is the series combination of the oxide capacitance and the quantum capacitance. Capacitance-voltage measurements of test structures with identical high-k deposition condition and thickness yielded a gate oxide capacitance of ~600 nF/cm^2, which resulted in an effective $C_{gc}\sim 124.5\ f_F$ for the device. Owing to source and drain metal contact resistance, the intrinsic device g_m is different from the measured extrinsic transconductance symbolized by $G_m(dI_d/dV_g)$. In brief, G_m includes the detrimental effects of metal resistance, while g_m is solely due to the graphene transistor. From classical circuit theory, the transistor with source resistance (R_S) and a possible drain resistance (R_D) is considered a source-degenerated common-source transistor [108]. The intrinsic g_m can be extracted from the extrinsic G_m via the relation for source-degenerated transistor

$$g_m = \frac{G_m}{1-G_m[R_S+(R_S+R_D)/(G_m G_{ds})]} \tag{11}$$

where G_{ds} is the extrinsic device conductance (dI_d/dV_d). Measurements of a test structure with the same source and drain metal patterns of the device yielded an estimate of 68 Ω and 22.67 Ω for R_S and R_D respectively. The experimental device was biased around the peak (hole) transconductance (~100 μS at $V_D=0.01$ V). Since the device is operating in the linear region, it is expected that the G_m scales linearly with V_D, resulting in $G_m \sim 5$ mS at the drain bias of 0.5 V used for the f_t measurement. From above relation, the intrinsic g_m is extracted to be 24.2 mS. This results in an estimated $f_t \sim 30.9$ GHz in close agreement with the experimental deembedded data shown in Figure 5.5. The extrinsic f_t can also be estimated analytically by including the parasitic overlap capacitances. The sum of the overlap of the source and drain fingers with the gate finger is about 400 nm, which results in an estimated extrinsic $f_t \sim 3$ GHz which is in good agreement with the experimental data reported in Figure 5.5. We note that the ratio between intrinsic and extrinsic cut-off frequency in this work (~a factor of 8) is significantly less than the ratio between intrinsic and extrinsic frequencies reported in other works, which could be as large as a factor of 128 [102.109].

As mentioned in the previous chapter, the reduction in the channel length is the most straightforward way to improve the intrinsic f_T performance without significant changes in device structures or deposited materials. Here, we scaled the channel length of the GFETs down to 0.25μm. Without any change in device structures, the intrinsic f_T has been improved by 40% from 25 GHz to 32 GHz as shown in Figure 5.7. The former device has the channel length of 0.5μm and the later device has the channel length of 0.25μm. Devices for both measurements are biased at the same high lateral electric field of 1 V/μm. Initially, the 2X improvement in the intrinsic transit frequency was expected due to the reduction in its channel length from 0.5μm to 0.25μm. However, due to the lower quality graphene film of the later device with low field-effect mobility results in relatively moderate improvement in the intrinsic transit frequency by 40%. Despite of this moderate improvement, further investigations to improve the quality of CVD-grown graphene films and more aggressive reduction in its channel length will invariably provide even higher performances suitable for future high-performance electronic applications. Also from the analysis on the extrinsic and intrinsic high-frequency performances, currently the overlap parasitic capacitance from the device structure limits the extrinsic device speed and highlights the need for a suitable self-aligned FET structure compatible with bendable flexible substrates.

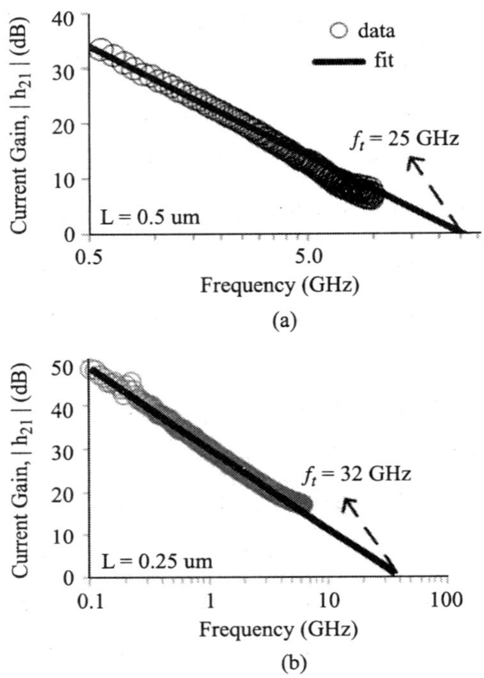

Figure 5.7. Channel-length scaling of GFETs.

(a) Intrinsic cut-off frequency measured from the 0.5μm channel-length device. V_{DS} = 0.6 V.

(b) Intrinsic cut-off frequency measured from the 0.25μm channel-length device. V_{DS} = 0.3 V.

Mechanical Bendability

The components integrated in the fabricated devices have two broad categories: flexible components and rigid components. While graphene film (with ϵ_{CR}~25%), and PI substrates (with ϵ_{CR}~10%), are considered as flexible components, high-k dielectrics (prepared by ALD) and metal electrodes are regarded as rigid components in this work [110-111]. While the maximum achievable mechanical flexibility (ϵ_{CR}~10%) is limited by the critical strain of PI substrates, the issue is how we can improve the flexibilities of rigid components. In order to understand the failure mechanisms of rigid components during mechanical bending, test structures with metal interconnects and metal-oxide-metal (MIM) capacitors wereprepared and experimentally evaluated as given in Figure 5.8. Figure 5.8(a) shows the change in normalized conductance from metal interconnects corresponding to the bending radius down to 0.7 mm; G stands for the conductance at the specific bending radius while Go represents the original value under the flat condition. Figure 5.8(b) presents the similar test results for capacitors; $10 \times 10\mu m^2$, $50 \times 50\mu m^2$, $100 \times 100\mu m^2$, and $200 \times 200\mu m^2$ unit devices were patterned on PI substrates; 30-nm thick ALD Al_2O_3 was deposited as dielectrics. The metal interconnects show lower conductances (40% reduction) at smaller bending radius owing to mechanical deformation and elongation making locally thinner lines with higher resistance. While supported on polymeric substrates, metal electrodes are uniformly stretched so as to approach their maximum allowable strain without degradation of over 4%, which corresponds to the bending radius of ~1.5 mm. Beyond this point, metal films show increasing resistance while still maintaining functionality, indicating good adhesion at the PI interface without de-bonding of metal films from the substrate [112]. In Figure 5.8(b), MIM capacitors are evaluated. Here, C and Co are the capacitance at the specific bending radius and the original capacitance underthe flat condition, respectively. The largest unit cell (CAP4 with $200\mu m \times 200\mu m$ dimension) shows dramatic reduction in its value at the radius of 1 mm and total failure at smaller radius. CAP2 (50μm × 50μm) and CAP3 (100μm × 100μm) cells failed at the minimum bending radius of 0.7 mm. The smallest unit cell (CAP1) didn't lose functionality even at the smallest bending radius. These results supports the idea that by predefining local dielectric islands where the active channel is located, the bendability of the flexible devices can be further improved and maximized with perhaps a small degradation of the electrical properties owing to increased electrode resistances.

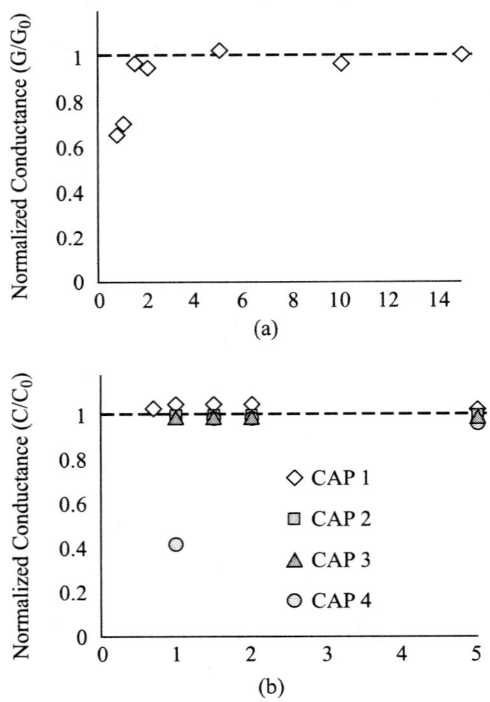

Figure 5.8. Mechanical bending experiments for metal interconnects and MIM capacitors on PI films.

(a) The change in conductance of metal lines.
(b) The change in normalized capacitances of MIM capacitors. CAP1, CAP2, CAP3, and CAP4 are 10 × 10μm², 50 × 50μm², 100 × 100μm², and 200 × 200μm², respectively. The dashed line is a visual guide indicating invariant properties over a wide bending radius.

The mechanical flexibility of the GFET was also evaluated with custom designed bending test fixtures. In order to achieve high flexibility, the rigid components (electrodes and gate dielectrics) integrated in the device are patterned. While un-patterned devices only survive the maximum bending radius of 2 mm, which corresponds to ~3% tensile strain mostly due to the dielectric breakdown, the patterned devices can survive bending down to a radius of 0.7 mm, which corresponds to ~8.6% strain. The tensile strain (ϵ) is computed according to the relation $\epsilon = t_s/(2R_B)$ where t_s and R_B are the substrate thickness and bending radius respectively [65]. During the measurements, the normalized resistance extracted from the device model has been monitored as shown in Figure 5.9(a); no noticeable degradation was observed. Figure 5.9(b) shows the robust electrical performance of the device; the mobility remains over 80% compared to the value at the flat condition at the minimum bending radius of 0.7 mm. Repeated measurements under the minimum bending radius shown in Figure 5.9(c) and 5.9(d) confirm the reliability of the graphene device structure. At lower bending radius, we observe mechanical breakdown of the gate dielectric leading to high gate leakage and poor gate modulation, however, the metal electrode is structurally intact and remains conductive [113-114]. The graphene transistors were also evaluated in a variety of harsh conditions as given in Figure 5.10, that can be hazardous for-conventional monolithic electronics. The sample was exposed to mechanical stresses such as a walking test at ~4 ft/sec, and automobile driving test at ~5 miles/hr. The human and automobile weigh approximately 115 lbs and 3300 lbs, respectively. As seen in Figure 5.10, the GFETs retain their electrical functionality under these harsh conditions. In the graphene device selected for monitoring, the degradation in current was less than 4%. The mobility dropped by as much as ~40% after immersion in warm coffee, but recovered to 90% of the initial mobility after the walking test, indicating that the impact of the warm fluid is at least partially reversible likely owing to thermal relaxation and fluid desorption.

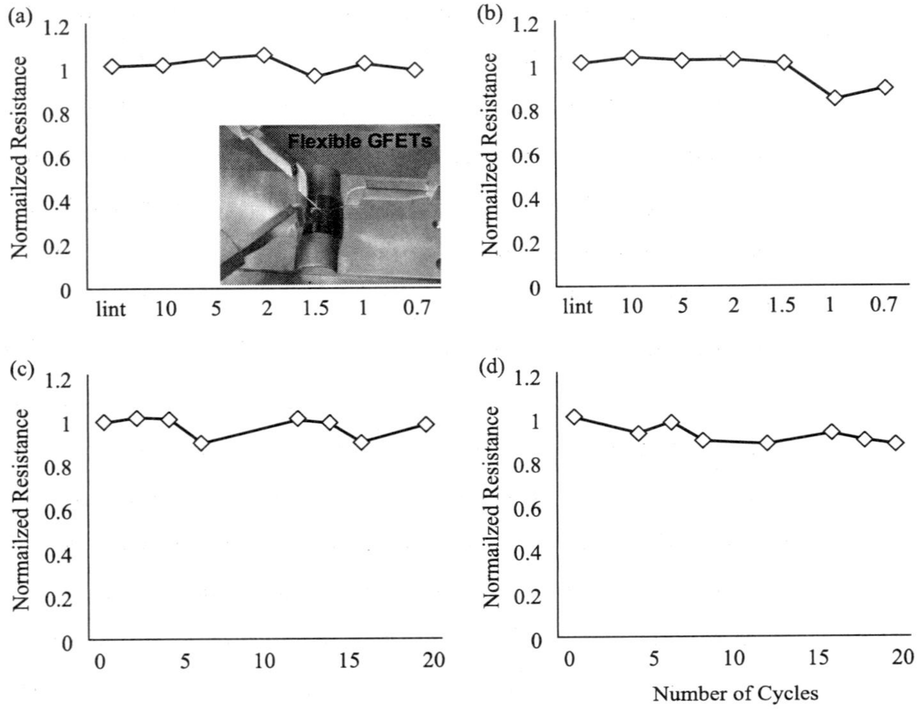

Figure 5.9. Mechanical bending measurements.

The change in (a) normalized resistance and (b) normalized mobility under different bending radii down to 0.7 mm. The inset in (a) shows the flexible GFETs attached to the bending fixture. Repeated measurements of (c) normalized resistance and (d) normalized mobility at the minimum bending radius of 0.7 mm.

These electrical results suggest that nitride passivated GFETs on PI provide a route for realizing high-performance rugged nanoelectronic systems. More systematic precision uniaxial compressive loading and impact testing are required for further electro-mechanical understanding. Figure 5.10(e) shows the frequency doubler performance of the device evaluated after completing dynamic loading tests. The frequency doubler evaluated in this work showed high-spectral purity ($>$ 90%) while affording a peak conversion loss of ~39.7 dB, which is within 5 dB of our earlier report achieved with graphene transistors on a smooth single-crystal quartz substrate [95]. These results motivate further developments of graphene-based analog and RF circuits on flexible sheets.

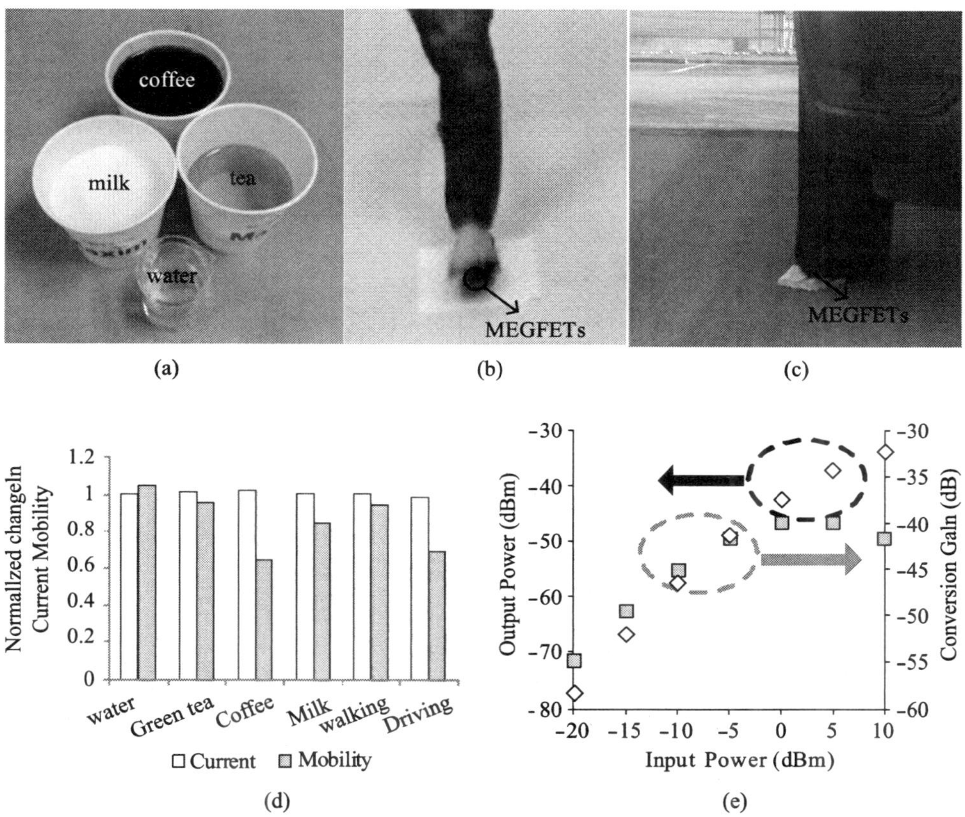

Figure 5.10. Experiments involving harsh conditions.

(a) Still image of the devices been walked over to emulate a slow-moving load.

(b) Still image of 2002 Honda CRV sport-utility vehicle going over the devices.

(c) Doubler characteristics. The output power at the doubled frequency (2 f) and the conversion gain as a function of the input power at the fundamental frequency (1 f = 2 MHz) are given. The gate is biased at the Dirac point and V_D = 300 mV.

(d) ???

(e) ???

Liquid Exposure

The effective encapsulation and protection of the flexible GFET by an hybrid bilayer (inorganic/organic) coating to liquid exposure (DI water) has been evaluated by the static contact angle measurements. The FTA200 contact angle goniometer is used to measure the contact angles for different passivation materials investigated including bare PI films, cured PI films, PECVD Si_3N_4, and CYTOP® coating on the flexible devices. The fabricated sample was self-passivated at the back-side by the bare PI substrate, while the top-side of the device was implemented on the smooth, cured PI surface and passivated by Si_3N_4/CYTOP bilayer.

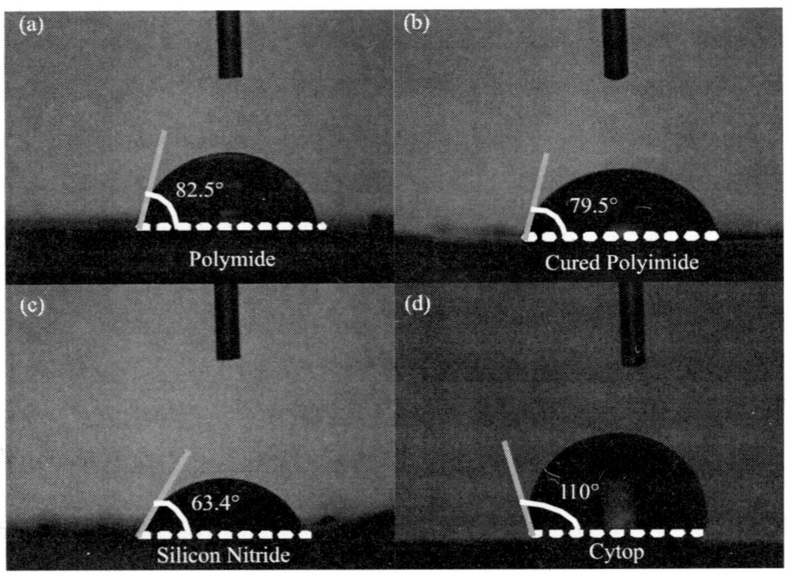

Figure 5.11. Contact angle measurements. 20μL DI water dropped on prepared samples including

(a) as-received commercial polyimide film
(b) cured polyimide film
(c) PECVD Si_3N_4
(d) Cytop coated on Si_3N_4/PI

Figure 5.11 shows the optical images captured by the optical microscope attached to the goniometer. During the measurement, DI water of 20μL is dropped on each surface. Bare PI filmsoffer high hydrophobicity with high contact angle of 82.5°, which affords good protection to DI exposure. However, the bare PI film is not recommended for integrated graphene devices due to the surface roughness and requires the coating/curing process to smoothen the surface. After curing the spin-coated smoothing layer, the hydrophobicity of the PI surface is slightly reduced down to 79.2°. Si_3N_4 doesn't improve the hydrophobicity, though, it provides outstanding mechanical robustness even to extreme static loading conditions. Of the different coating films investigated in this work, we identified highly hydrophobic CYTOP® which offers a high contact angle exceeding 110° as the most suitable water protection layer.

Figure 5.12 present the electrical measurements under DI water exposure; Figure 5.12(a) and 5.12(b) are the data from Si_3N_4 passivated devices and Figure 5.12(c) and Figure 5.12(d) are the data from the bilayer (Si_3N_4/ CYTOP) passivated devices. The devices were dipped into DI water for 10 s, dried with N_2 flow, and repeatedly measured. To evaluate the short-term protection against DI waterimmersion, the measurements were repeated up to 150 s while monitoring the changes in the normalized resistance and the carrier mobility. As for the long-term DI water exposure, the same measurements were performed after immersion for 2 days. The dotted lines in Figure 5.12 represent the virtual boundary between short-term measurements and long-term measurements. The normalized resistance does not show any noticeable degradation for both cases, while the carrier mobility degraded significantly for nitride passivated devices down to 30% of the initial value. This reduction in the mobility is due to nonideal water protection from the nitride passivation layer. A low contact angle (less hydrophobic surface) of the layer allows water molecules to wet the surface and eventually introduces carrier traps in the dielectric interfaces, hence, weakening the gate modulation. The bilayer coated device shows no degradation of its resistance and slight degradation of its mobility, while providing 80% of the initial electrical performance. The improved water-resistant protection is attributed to the strong hydrophobic surface of CYTOP. This result suggests that functional coatings can be employed to protect flexible GFETs from water or liquid exposure, which is anticipated as a future requirement for robust flexible smart systems.

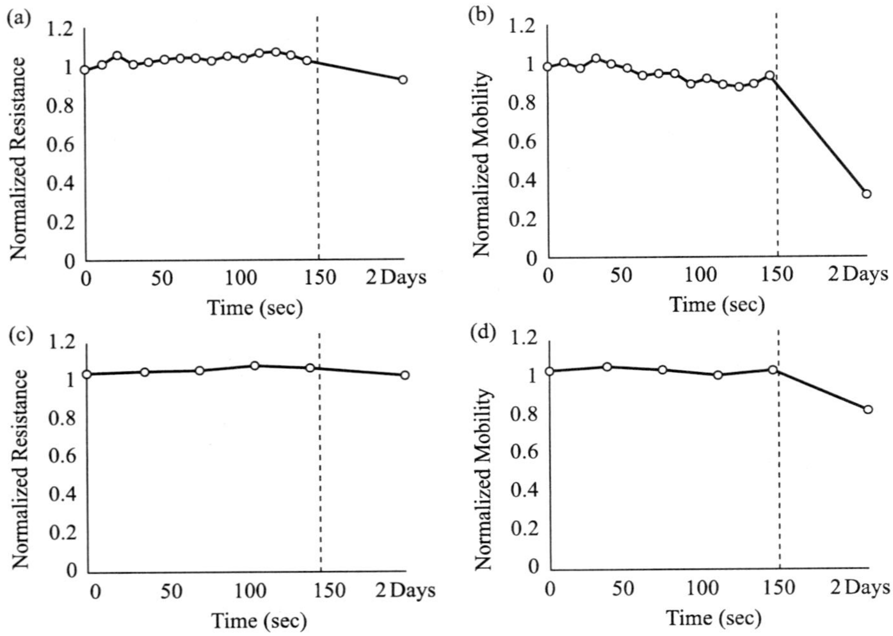

Figure 5.12. Immersion test results for the normalized resistance and mobility. Data shown on the left side of the dashed line are for short-term exposure to DI water, and data on the right side of the line are for long-term exposure to DI water.

(a), (b) Normalized resistance and mobility measured from Si_3N_4 passivated devices.

(c), (d) Normalized resistance and mobility measured from Cytop/Si_3N_4 passivated devices.

Chap 6

Transformation of the Electrical Characteristics in GFETs

- Chemical-vapor-deposited mono-layered graphene
- Hybrid graphene/organic semiconductor FETs
- Transformation of the electrical characteristics of GFETs with fluoropolymer

Electronic devices with an active channel consisting of a single layer graphene have been of great scientific and technological interest due to its high carrier mobility and stable mechanical and chemical properties [1-7]. In particular, very fast charge carrier transport combined with mechanical and thermodynamic stability can be key enablers to promising nano-scale and sub-millimeter wave electronic devices applications. Graphene based FETs have been shown to operate at very high frequencies. However, since a single layer graphene has no bandgap in principle, the current in graphene FETs is difficult to turn off completely by gate bias compared to silicon-based devices, which limits the possible use of such transistors in digital circuits. Formany electronic circuit applications it will be desirable to improve the on-off current ratio so that graphene resembles a semiconductor more than a semimetal. Research and manufacturing of organic semiconductors has been developing rapidly during the past few decades. Most of the studies on graphene in organic devices have focused on its metallic conductivity for use as an electrode. It must be noted that there have been no studies on FETs with hybrid graphene/organic semiconductor active layers. In this chapter, we describe the electrical characteristics of hybrid graphene/organic semiconductor FETs employing a-sexithiophene (α-6T) and hexadecafluorocopperphthallocyanine ($F_{16}CuPc$) for the organic materials. We also demonstrate a general method to favorably transform the electrical characteristics of graphene FETs by capping with fluoropolymers such as CYTOP®. Key metrics such as mobility, on-off current ratio and residual carrier density, no, are all improved. We present a hypothesis as to why this alteration happens.

Chemical-vapor-deposited mono-layered graphene

As a first step, good quality mono-layered graphene films were synthesized on 500 nm thick e-beam evaporated copper films grown on a Si/SiO_2 substrates, by the low-pressure chemical vapor deposition (LPCVD) as described below. The samples were first annealed for 5 minutes at 1000 ℃ in a hydrogen-containing ambient. The hydrogen gas was purged away at the end of the annealing process and ultra-high purity (99.99%) methane was circulated at a flow rate of 10 sccm for 5 minutes during the growth process. After growth, the chamber was slowly cooled down to below 180 ℃ before unloading of samples.

The graphene films were then transferred to a Si/SiO_2 substrate by a conventional wet-transfer process. In the wet-transfer process, graphene on copper film was firstspin coated with poly (methyl methacrylate) (PMMA) at 4000 rpm for 1 minute. The sample was kept in desicator overnight for removing the solvent (chlorobenzene) in PMMA. It was then placed into Buffer Oxide Etch (6 : 1) to etch way the SiO_2 to have PMMA-Graphene-Cu film detached from the substrate. It is important to remove the SiO_2 first to have the surface of copper film exposed. Otherwise we have to use strong solution of copper etchant to etch it from sides and it causes defect in graphene which shows up in the Raman spectrum after transfer. The film was carefully taken out of the Buffer Oxide Etch with a spoon and put in DI water. It was then placed in dilute ammonium persulfate solution (Transene, APS 100) to etch away the copper. After the copper film was completely etched away, the PMMA-Graphene film was rinsed with DI water couple of times. The target substrate was held inside the water in an inclined manner and was positioned just under the film. The substrate was then gently pulled out of the water with the PMMA film on top. After the sample was dried in desicator overnight to remove water, it was heated on hotplate at 120 ℃ for 2 minutes. This step is crucial because it involves some tradeoff. Heating PMMA to higher temperature (usually 180 °C, which is above PMMA glass transition temperature-Tg) for longer time causes the PMMA to reflow and film adhesion to the substrate to be better. However, this also causes the PMMA to stick to graphene and it gets difficult to remove it later in acetone without appropriate annealing condition. This is why the sample was heated just above PMMA Tg for a short time. PMMA was then removed in acetone. The acetone was replaced with fresh supply twice over the span of 24 hours. Finally the sample was rinsed in IPA and dried in air. Figure 6.1 shows Raman spectra on CVD grown mono-layered graphene. The full width at half maximum (FWHM) of the 2D-peak is ~29 cm^{-1}. The intensity ratio between 2D and G peaks is 2.8. The D peak (defect related peak) intensity is small or negligible. AFM images of the graphene show a polycrystalline film, as shown in Figure 6.1. These results indicate that the mono-layered graphene in the FET is of high material quality.

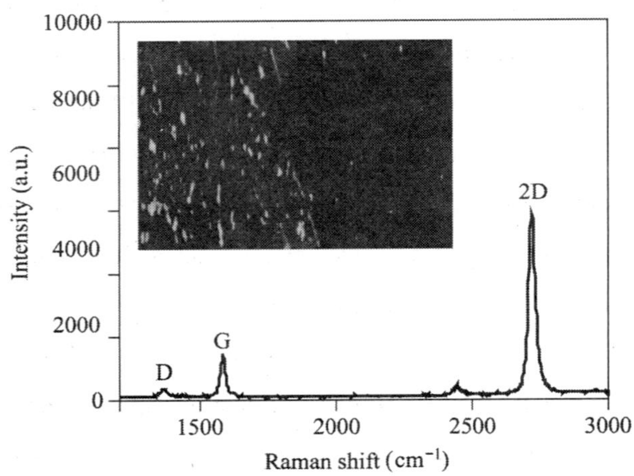

Figure 6.1 Raman spectra on CVD grown mono-layered graphene with a 442 nm blue laser. The inset shows AFM image of CVD grown mono-layered graphene after graphene transfer

Hybrid graphene/organic semiconductor FETs

Figure 6.2 (a) and (b) show the schematic cross-section and optical image of a hybrid graphene/organic semiconductor FET. The device fabrication process starts with 1 − 10 Ω-cm n-doped silicon substrate employed as the bottom-gate electrode. It is thermally oxidized to result in a 290 nm thick silicon dioxide bottom gate insulator. A single layer graphene was synthesized by a LPCVD on a 500 nm thick copper film. Source/drain electrodes were patterned by electron-beam lithography and lift-off and then a chrome/gold (2.5 nm/50 nm) bi-layer was deposited by thermal evaporation. The sample was kept in high vacuum for 2 days to minimize impurity incorporation. A 15nm thick layer of organic semiconductor (either α-6T or $F_{16}CuPc$) was deposited

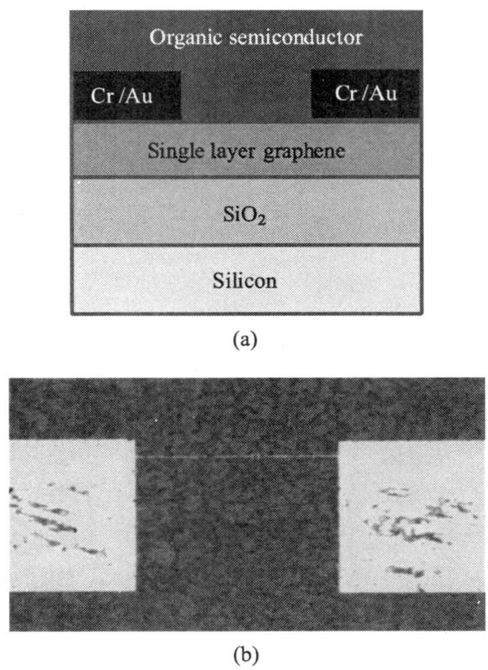

Figure 6.2 (a) schematic cross-section and
(b) optical image of a hybrid graphene/organic semiconductor FET

by thermal sublimation at a rate of 0.1 Å/s under a pressure of 10^{-7} Torr. Substrate heating at 180 ~ 200 ℃ was applied to improve the quality of semiconductor films. Optimized bottom gate bottom contact α-6T and $F_{16}CuPc$ FETs possess field-effect mobilites of 0.03 and 0.02 cm^2/V-s in air, respectively. These values are consistent with the highest reported mobilities for these materials [115]. In order to remove organic semiconductors from hybrid graphene/organic semiconductor FETs, they were subject to heat treatment using a substrate heater at 250 ~ 300 ℃ under high vacuum. This causes the organic semiconductor film to sublime away. Hybridgraphene/organic semiconductor FET possesses a channel width of 8μm and a channel length of 4μm. DC measurements for the device characteristics were carried out using Agilent 4155C semiconductor parameter analyzer.

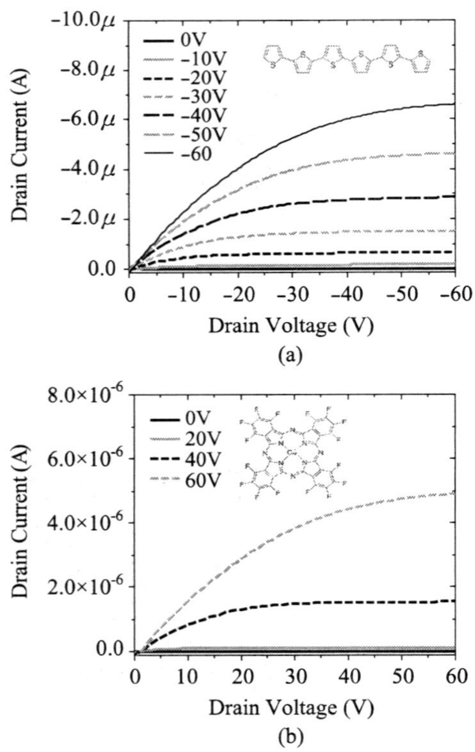

Figure 6.3 (a) The output characteristics of optimized bottom-gate bottom-contact α-6T and $F_{16}CuPc$ FETs

Figure 6.3 (a) and (b) show the output characteristics of optimized bottom-gate bottom-contact α-6T and $F_{16}CuPc$ FETs. α-6T FETs possess hole-dominant transport whereas $F_{16}CuPc$ exhibits electron-dominant transport. In order to improve the quality of p-conjugated organic semiconductor, substrate heating was employed. Good output characteristics were observed with low electrical contact resistance. The extracted field-effect mobilities from α-6T and $F_{16}CuPc$ FETs are 0.03 cm^2/V-s and 0.02 cm^2/V-s, respectively.

Figure 6.4 (a) shows the transfer characteristics of a hybrid graphene/organic semiconductor FET employing α-6Tsemiconductor at the drain-source voltage of 0.1 V during the sweep of gate voltage from -70 V to 70 V. The transfer characteristics of a graphene FET are not degraded by depositing organic semiconductor onto the graphene layer. It can be noted that the transfer characteristics of hybrid graphene/organic semiconductor FETs return to its initial state (i.e., before organic semiconductor deposition) after removing α-6T semiconductor layer. The electronic properties of graphene can be tuned favorably using the capping layer of organic semiconductor. In order to extract the key parameters, a diffusive transport model based on total resistance of the graphene device was used . Figure 6.4(b) shows there is good agreement between the experimental data and diffusive transport model. Optimized graphene/organic semiconductor FETs with α-6T possess field-effect mobility of 1289 cm^2/V-s in air, which is slightly improved from 1176 cm^2/V-s in single layer graphene FET without a decrease in the on-off current ratio. The extracted no is 10^{12} cm^{-2} and width-normalized contact resistance is $\sim$1.9 kΩ-μm. Furthermore, removal of the organic semiconductor layer from graphene FETs results in a return to the original electronic properties. The results suggest that there are weak reversible electronic interactions between organic semiconductor and graphene layers.

The on-off current ratio and the field-effect mobility in a graphene FET are typically decreased after depositing most additional layers on grapheme [116].

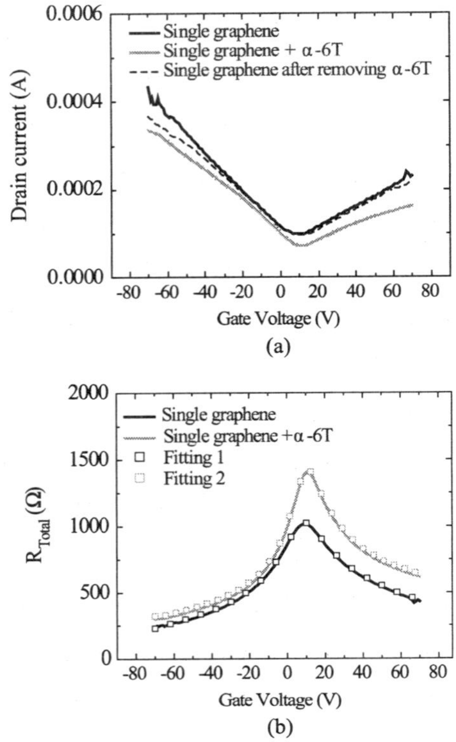

Figure 6.4 (a) The transfer characteristics of single layer graphene, graphene by capping with α-6T and graphene FETs after removing α-6T at the drain-source voltage of 0.1 V during the sweep of gate bias from -70V to 70V. The inset shows the chemical structure of α-6T and (b) the total resistance along with diffusive transport modeling fit

Figure 6.5 shows the transfer characteristics and the normalized resistance of single layer graphene by capping with α-6T semiconductor layer and with silicon dioxide. Compared to silicon dioxide on a graphene FET, the on-state current ratio in graphene/α-6T FET was improved from 3 to 5 and on-current was increased by factor of two without degradation in field-effect mobility. It means that weak interaction between α-6T semiconductor and graphene influences the electronic characteristics including charge carrier transport, which is obviously distinguished from silicon dioxide. The nature of this interaction is elucidated later in this chapter.

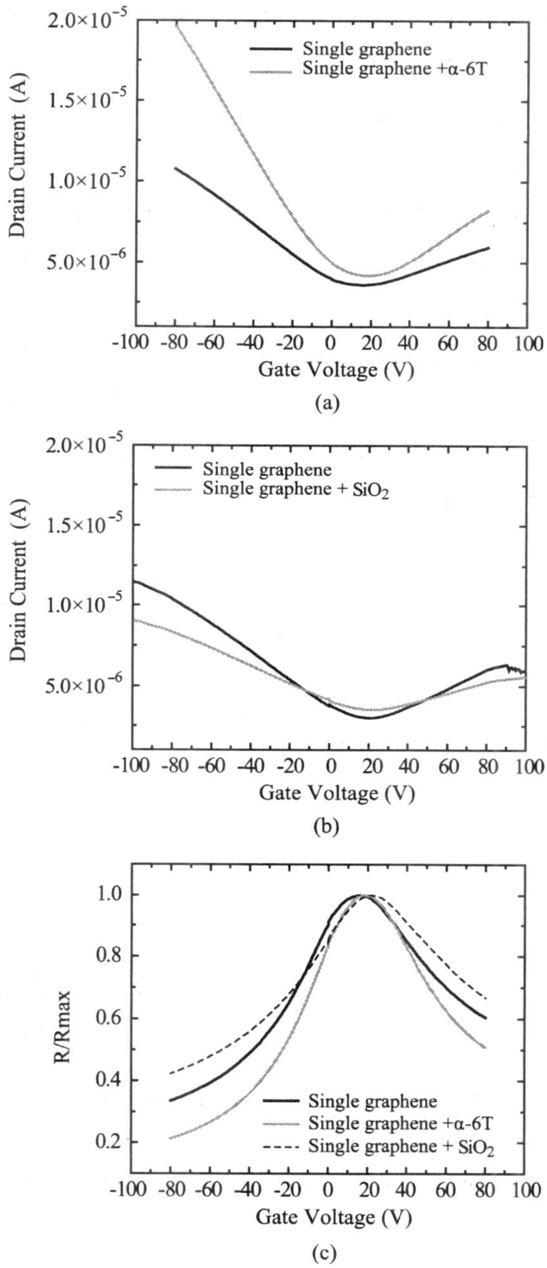

Figure 6.5 (a) The transfer characteristics of single layer graphene by capping with (a) α-6T and
(b) silicon dioxide.
(c) the normalized resistance in single layer graphene and graphene FETs by capping with α-6T and silicon dioxide

It has been reported that fluorination of graphene can dramatically alter its electronic and optical properties, affording orders of magnitude increase in the resistivity and opening of a bandgap of several eV [117]. The effect of capping with $F_{16}CuPc$ on graphene FETs was investigated, as shown in Figure 6.6. Compared to the plain single layer graphene FET, capping with $F_{16}CuPc$ improved on-off current ratio from 3 to 5 as well as the field-effect mobility from 1292 to 1367 $cm^2/V\text{-}s$ in air. Furthermore, the Dirac voltage point was shifted to a more negative voltage. The value of the extracted no is decreased from 1.3×10^{12} to $1.0 \times 10^{12}\ cm^{-2}$, indicating that a significant favorable modification of electronic properties occurs with the introduction of the fluorinated semiconductor.

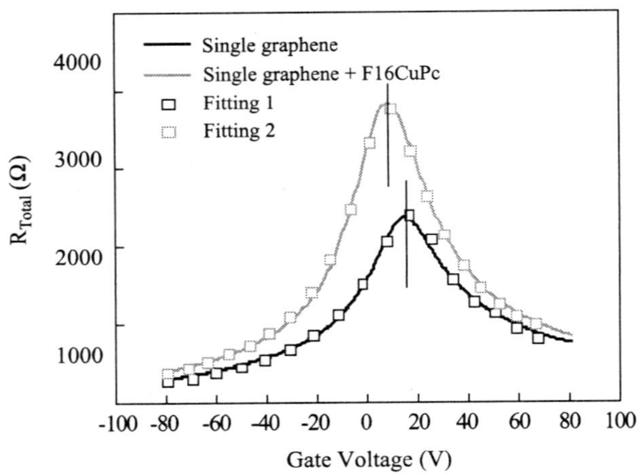

Figure 6.6 (a) The total resistance in single layer graphene and graphene FETs by capping with $F_{16}CuPc$ along with diffusive transport modeling fit The inset shows the chemical structure of $F_{16}CuPc$

Transformation of the electrical characteristics of GFETs with fluoropolymer

For many electronic circuit applications it will be desirable to improve the on-off current ratio so that graphene resembles a semiconductor morethan a semimetal. In achieving such a transformation, the high mobility must not be reduced and preferably increased. We demonstrate another method to favorably transform the electrical characteristics of graphene FETs by capping with fluoropolymers such as CYTOP® and Teflon-AF®. With this approach we get even better results compared to that obtained with organic semiconductors described in the last section. The conductivity at the Dirac point is reduced and the mobility is increased, leading to an improvement in the on-off current ratio to ~10, the highest that has been hitherto achieved in graphene. Remarkably, all key graphene device metrics are improved including electron-hole transport symmetry, Dirac voltage, and reduced impurity doping. We note that, in general, attempts to coat graphene with inorganic dielectrics such as silicon dioxide and aluminum oxide have not resulted in an improvement in electrical characteristics. Importantly, these results have been achieved in graphene grown by wafer-scale CVD process. From a practical standpoint, this is a significant advance in that it offers a clear path to improve the performance characteristics of graphene FETs in which the active material is grown by wafer-scale CVD. CVD graphene is the most promising method of realizing large area graphene and in adapting graphene for silicon CMOS and flexible electronics [118]. The improved mobilities and reduced conductivities will open up the range of applications for graphene circuitry. Figure 6.7 shows the illustration of fabrication process flow of a mono-layered graphene FET after capping with the fluoropolymer. Good quality mono-layered graphene films were formed as described in section 6.2. Oxygen plasma reactive-ion-etching (RIE) was used to pattern the active channel region, to remove the superfluous graphene and to ensure device isolation. Source/drain electrodes were

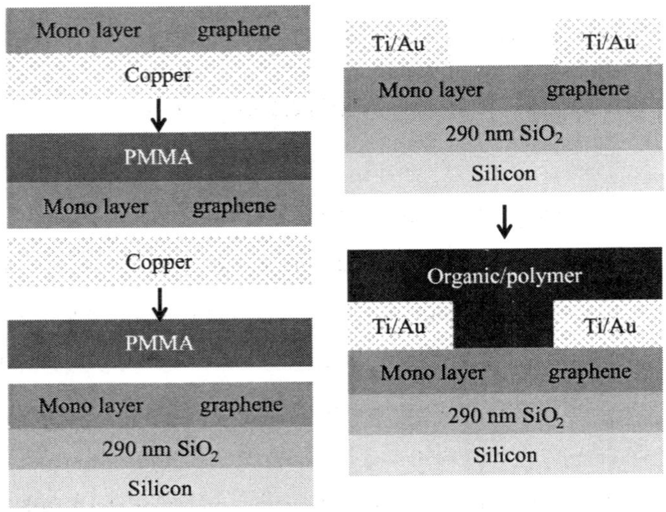

Figure 6.7 Illustration of fabrication process flow of a mono-layered graphene FET after capping with the fluoropolymer

patterned by electron-beam lithography and lift-off. The titanium/gold (2.0 nm/50 nm) bi-layers that form the source/drain contacts were deposited by thermal evaporation under high vacuum. Deposition of good quality titanium and gold in high vacuum conditions ($\sim 10^{-7}$ torr) are key to realizing low-resistance electrical contacts. The samples were kept in high vacuum for 2 days to minimize impurity incorporation. Mono-layered graphene FETs fabricated as described above feature possess a channel width of $5\mu m$ and a channel length of $1\mu m$. A 90 nm thick layer of the fluoropolymer, CYTOP® (Asahi Glass Co.) was deposited by spin-coating a diluted CYTOP® solution (CYTOP®: solvent = 1 : 10) on mono-layered graphene and annealed gradually from 30 ℃ to 180 ℃ for over a span of 1 hour in a nitrogen atmosphere. A 140 nm thick layer of Teflon-AF® (Dupont Co.) was also spin-coated with as-supplied Teflon-AF® solution on mono-layered graphene. The samples were annealed gradually from 30 ℃ to 300 ℃ for over a span of 1 hour in a nitrogen atmosphere. In order to remove the CYTOP® from graphene FETs, the samples were immersed in a CYTOP® solvent for 24 hours. This removes most of the CYTOP®. The temperature-dependent measurements were performed in a Desert Cyrogenics vacuum probe station with a chamber pressure lower than 10^{-3} Torr. DC measurements for the device characteristics were carried out using Agilent 4155C semiconductor parameter analyzer.

Figure 6.8 (a) shows the transfer characteristics of as-deposited mono-layered graphene FET without capping with CYTOP®, with capping with CYTOP®, and after removal of CYTOP®. The drain-source voltage is 0.1 V

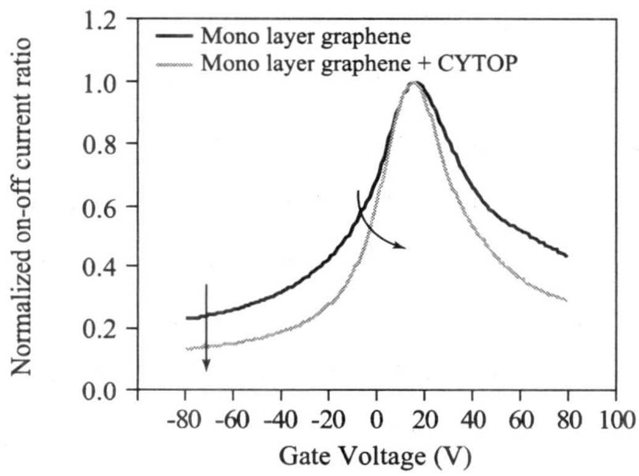

(a)

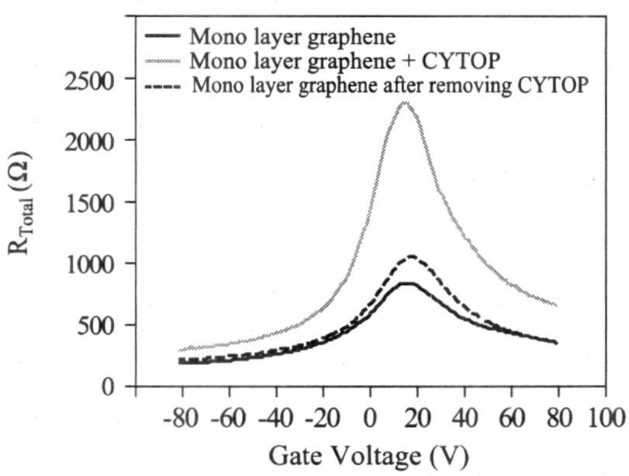

(b)

Figure 6.8 (a) The transfer characteristics of mono-layered graphene, graphene with CYTOP® capping layer and graphene FETs after removing CYTOP® .The drain-source voltage is 0.1 V and
(b) Improvement in the normalized on-off current ratio after depositing CYTOP® on mono-layered graphene FET.

during the sweep of the gate voltage from −70 V to 80 V. It is generally observed that device characteristics such as on-state current, field-effect mobility and on-off current ratio in graphene FETs are reduced after the deposition of most dielectrics materials on graphene due to charge scattering [119-120]. It is observed that in the present case, the off-state current (at the Dirac point) is decreased substantially, resulting in a net improvement in the on-off current ratio with the use of CYTOP®. This is illustrated more clearly in Figure 6.8(b) which shows that the on-off current ratio is improved from 5 to 10 after depositing CYTOP® on graphene FETs. In addition, the field-effect mobility was improved from 1731 to 3606 cm^2/V-s, width-normalized contact resistance is not appreciably altered from ~400 Ω-μm and residual carrier density no is reduced. The results indicate that electrical device characteristics of graphene FETs are significantly improved through the interaction between fluoropolymer and mono-layered graphene, which is different from the interaction of graphene with most dielectrics such as silicon dioxide, aluminum oxide and hafnium oxide. When the CYTOP® layer was removed by using CYTOP® solvent from mono-layered graphene FETs, the transfer characteristic tends to return to its initial state (i.e. that of mono-layered graphene before CYTOP® deposition), as shown in Figure 6.8(a). These results mean that the interaction between graphene and CYTOP® is weak enough to permit easy removal of CYTOP®.

Raman spectra were measured with a 442 nm blue laser on mono-layered graphene capped with CYTOP®. Figure 6.9 shows the change in the Raman spectrum of mono-layered graphene produced by the capping layer of CYTOP®. The intensity of the D band at 1350 cm^{-1} was strongly increased after capping the graphene with CYTOP®. At the same time, the intensity of the 2D band at 2700 cm^{-1} and the G band at 1580 cm^{-1} were slightly decreased. Significantly, a new low-intensity band at 2940 cm^{-1} was observed on mono-layered graphene FETs with CYTOP®, which is not observed before depositing CYTOP®. Appearance of this new band in the Raman spectrum was previously reported in fluorinated graphene.

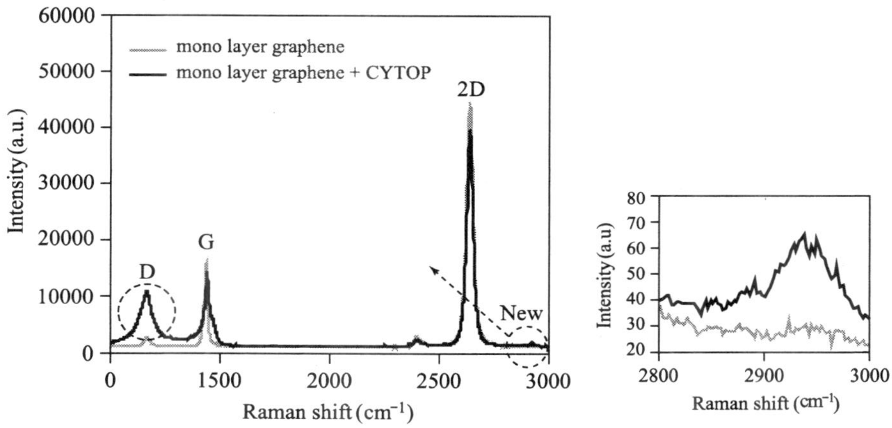

Figure 6.9 Change in Raman spectrum measured with a 442 nm blue laser of mono-layered graphene with and without CYTOP® capping layer

The C-F bonds in the CYTOP® structure are very polar and graphene is a highly polarizable material. The electronic interaction between the dipoles in the CYTOP® and the graphene can modify its electronic properties significantly, leading to the effects that we observe. We note that with non-polar organic capping layers such as pentacene, no significant changes in the electrical properties of graphene FETs are observed as shown in Figure 6.10.

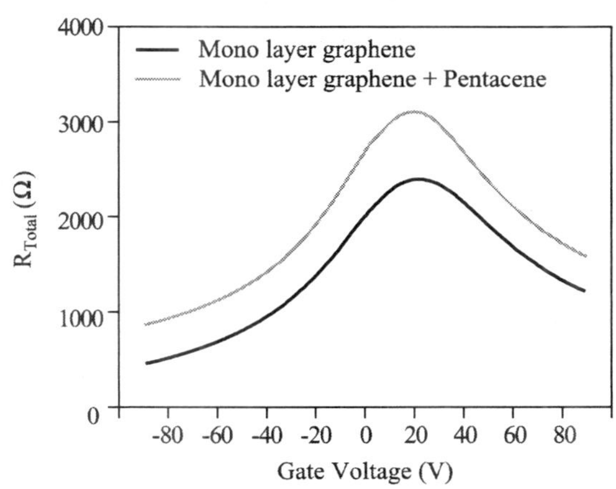

Figure 6.10 The total resistance of mono layer graphene with and without capping with Pentacene. There is no significant change in the on-off current ratio.

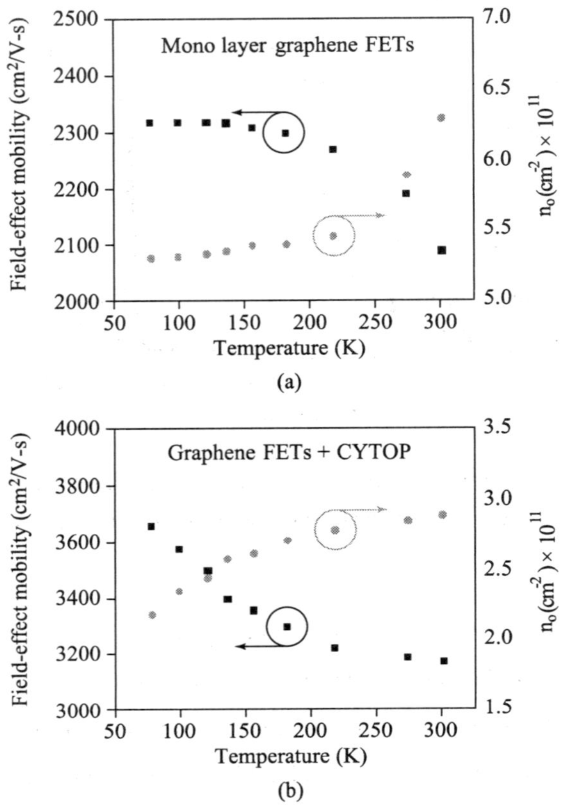

Figure 6.11 (a) Temperature dependent mobility and no in (a) mono-layered graphene FETs vs. (b) graphene FETs capped with CYTOP®: Studies show that with capping, the impurity scattering limited mobility continues to increase with reducing temperature which is accompanied by a reduction in no

We have verified our hypothesis in the case of a second fluoropolymer, Teflon-AF® possessing polar C-F bonds, which also results in a marked improvement in electrical properties of graphene upon capping. The origin of this improvement is in the strongly polar nature of the C-F chemical bond found in the capping materials we have employed together with the tendency of these materials to self-organize upon heat treatment such that there is an oriented layer of dipolar C-F bonds at the interface with graphene. Temperature dependent mobility and impurity studies also show that with capping, the impurity scattering limited mobility continues to increase with reducing temperature which is accompanied by a reduction in n_0 as shown in Figure 6.11.

The effect of capping with CYTOP® on doped mono-layered graphene FETs is shown in Figure 6.12. The on-off current ratio was improved from 6 to 9 as well as the field-effect mobility was increased from 1753 to 3045 $cm^2/V\text{-}s$ without change in width-normalized contact resistance after depositing

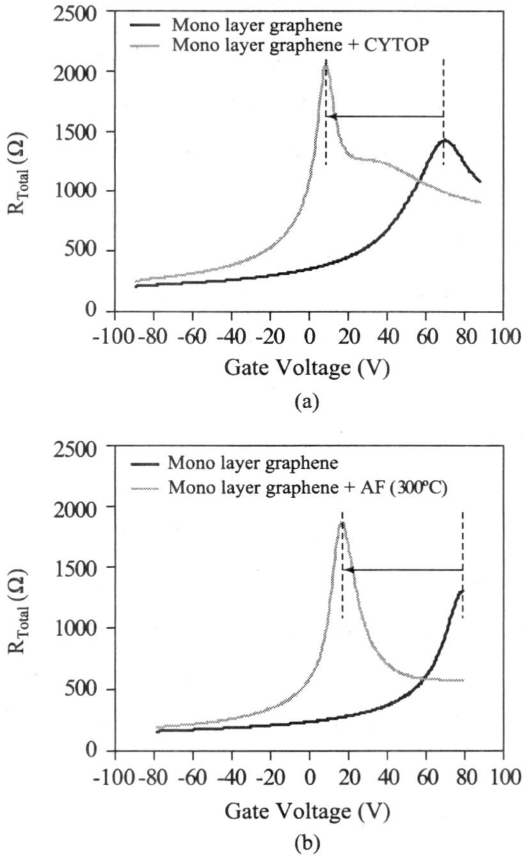

Figure 6.12 Transformation of characteristics of mono-layered graphene by capping with (a) CYTOP® and (b) Teflon-AF® at the drain-source voltage of 0.1 V during the sweep of gate bias from −80 V to 80 V. The substantial shift in the Dirac voltage toward zero after depositing CYTOP® and Teflon-AF® can be noted.

CYTOP®. In addition, the residual carrier density no is reduced from 1.03×10^{12} to 5.2×10^{11} cm^{-2}. Significantly, there is a large shift in the Dirac voltage point. The fluorocarbon capping method is therefore a way to restore or greatly improve the properties of graphene that are otherwise non-ideal. These effects are accompanied by a favorable shift in the Dirac voltage toward zero. In some cases of Teflon-AF®, this shift in Dirac voltage is very dramatic, with shift magnitudes in excess of 60 V, as shown in Figure 6.12.

The capping materials that we have chosen all possess C-F bonds and processing conditions have been employed that results in the materials ordering in a manner that results in a strong net dipole moment at the interface with graphene. Additional experiment and theoretical work is being done to understand these effects in more detail and will be reported elsewhere. The fact that the strength of this interaction is dependent on the annealing temperature of the CYTOP® and Teflon-AF® also suggests that annealing improves the ordering of the fluoropolymer and consequently the total dipole strength. In the case of the two fluoropolymers post-deposition annealing (at temperatures up to 300 °C) was employed to enable material reorganization and the side chain alignment that is commonly observed in these materials. For the organic semiconductor, $F_{16}CuPc$, material deposition was performed at elevated temperatures (up to 200 °C). Previous work has shown that such elevated deposition temperatures results in the best ordered materials resulting in relatively high mobility in these semiconducting films. It can be clearly observed as shown in Figure 6.13 that the electrical properties of CYTOP-coated graphene steadily improve with annealing temperature (in the range 60～180 °C) demonstrating that the side-chain alignment that accompanies such annealing is a key factor in the transformation of electrical properties. It is noteworthy that if the CYTOP® is removed after the annealing, then the original graphene characteristics are recovered indicating reversible non-covalent interactions.

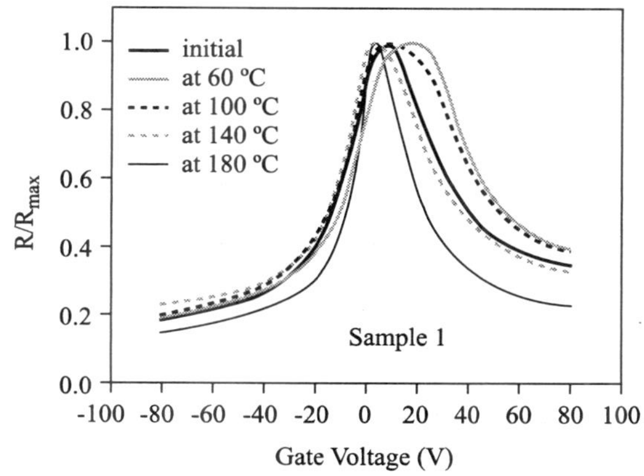

Figure 6.13 The electrical properties of CYTOP-coated graphene FETs with annealing temperatures

While many of these results can be explained in terms of a modification of the fine structure constant, we note that graphene is very polarizable and changes to the electronic structure and, consequently transport properties, can result from having oriented dipoles to top of the graphene. Additionally, we have modified the electronic environment on only one side of the graphene mono-layered. If both interfaces have highly polar bonds juxtaposed, then the favorable effects we report can be further enhanced.

Figures 6.14 (a) and (b) show the electrical characteristics of the GFETs before and after employing the CYTOP®, demostrating favorable positive as well as negative shifts in the Dirac voltage toward zero. Before coating, the GFETs showed a positive Dirac voltage of 25 V, and a negative Dirac voltage of −25 V due to the undesirable chemical adsorption during the fabrication processes. Since it is very difficult to avoid the incorporation of impuritiesthat attach to the graphene surface, large variation in the electrical characteristics including the Dirac voltage between samples (or between batches) has been observed. Generally, the device characteristics, such as on-state current, field-effect mobility and on-off current ratio in the GFETs were degraded after the deposition of most of the dielectric materials on graphene due to charge scattering. However, in the present case for fluoropolymer encapsulation, the resistance at the Dirac point substantially increased, thereby resulting in an improvement in the on-off current ratio, without a change in the electrical contact resistance. Upon coating with the appropriate fluoropolymer and optimization via thermal annealing, such variation was considerably reduced with the Dirac voltage shifting in both directions toward zero, regardless of its initial positions of the Dirac voltage. The results are difficult to achieve only through molecular chemical doping from electron-donating and -withdrawing functional groups, which doping exhibits uni-directional shift in the Dirac voltage. In addition, the electron and hole transport properties became more symmetric with mobilities of 4987 $cm^2/V\text{-}s$ for the holes, and 4628 $cm^2/V\text{-}s$ for the electrons. The symmetrical improvement in both the electron and hole transport is also difficult to explain to be a result of the charge transfer via electron-donating and electron-withdrawing from doping media into the graphene, where molecular adsorption of the functional groups shows exclusive improvement in either the electrons or holes. Such an improvement was observed in 30 samples fabricated in different batches at different times. We postulate that these effects were achieved by the carbon-fluorine dipole interaction between graphene and fluoropolymer that modifies the scattering effect of locally distributed impurities, thereby resulting in a modification of the electronic and material properties of the GFETs. The favorable shift in the Dirac voltage suggests that the fluoropolymer coating can help to relax the effects that electron-hole puddles in the graphene films have on charge transport. The presence of electron-hole puddles leads to a fluctuation of locally distributed chemical potential, and they also serve as long-range scattering center, which is significant for the low energy regime near the Dirac point. The bath with the strong dipole moments in the fluoropolymer seems to provide a charge buffer for the graphene layer, and accordingly, the nature of the impurity scattering could be modified.. When the CYTOP® layer was removed from the GFETs by using CYTOP® solvent, the device characteristic have a tendency to return to the initial state in both cases (i.e. to that of monolayer graphene before CYTOP® deposition), as shown in Figure 6.14. These results indicate that the electronic properties of graphene can be favorably tuned by using an interacting layer of a material such as CYTOP® that functions through reversible, non-covalent interactions.

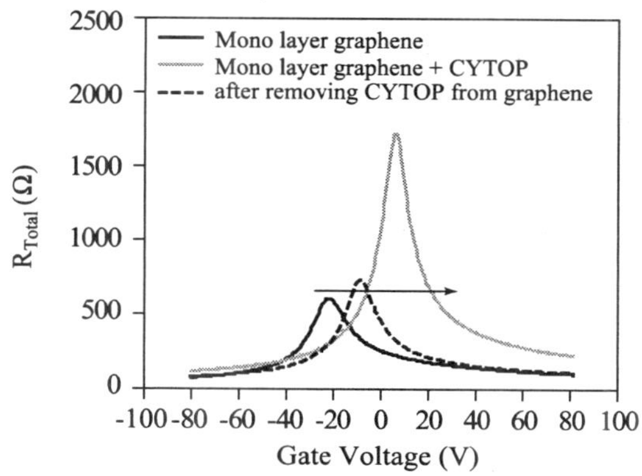

(a)

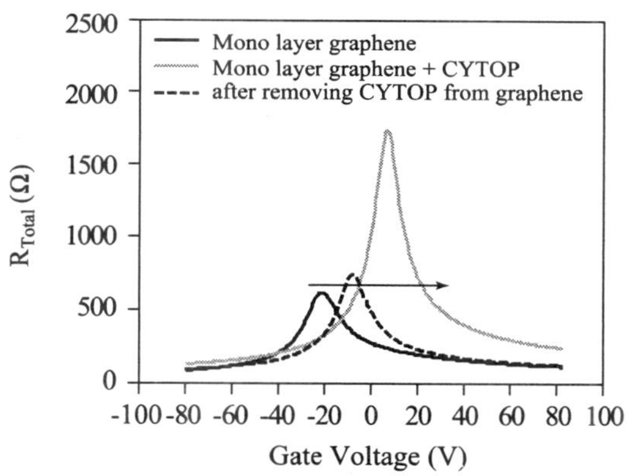

(b)

Figure 6.14 Transfer characteristics of GFETs before and after encapsulating with CYTOP® as an interacting layer and GFET after removing CYTOP® : with (a) negatively shifted and (b) positivelyshifted Dirac voltage

In addition to the modification of the local dielectric environment adjacent to the graphene layer, we investigated how the hydrophobic characteristics of the fluoropolymer films affects the long-term stability in terms of the device performance of the CVD grown monolayer GFETs. The hydrophobic properties of the surfaces of the fluoropolymer were verified by measuring a contact angle of 110° with an optical microscope attached to a goniometer. In contrast, the graphene surface had a contact angle of 91°. Figure 6.15 shows the variation in the field-effect mobility with the exposure time to air in the GFETs, with and without capping with CYTOP®. Less degradation in the electrical characteristics can be observed when the graphene is encapsulated with the hydrophobic fluoropolymer, which results from repelling the native hydroxyl (−OH) functional groups from the graphene surface, and preventing a direct interaction between graphene and water molecules from the outside environment.

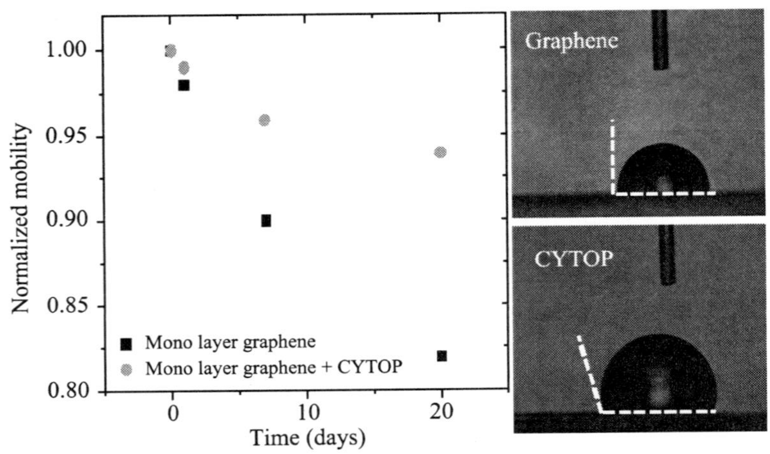

Figure 6.15 Variation in field-effect mobility with exposure time to air in CVD GFETs with and without encapsulating with CYTOP® and the contact angle of graphene and CYTOP® measured by the optical microscope attached to the goniometer

(P(VDF-TrFE)) has been used to fabricate a non-volatile ferroelectric gating devices by employing polymer piezoelectric materials that possess a strong dipole-dipole interaction. As mentioned above, the additional charge scattering that is induced by a polycrystalline P(VDF-TrFE) film used as an encapsulation layer may degrade the electrical characteristics of the GFETs. However,

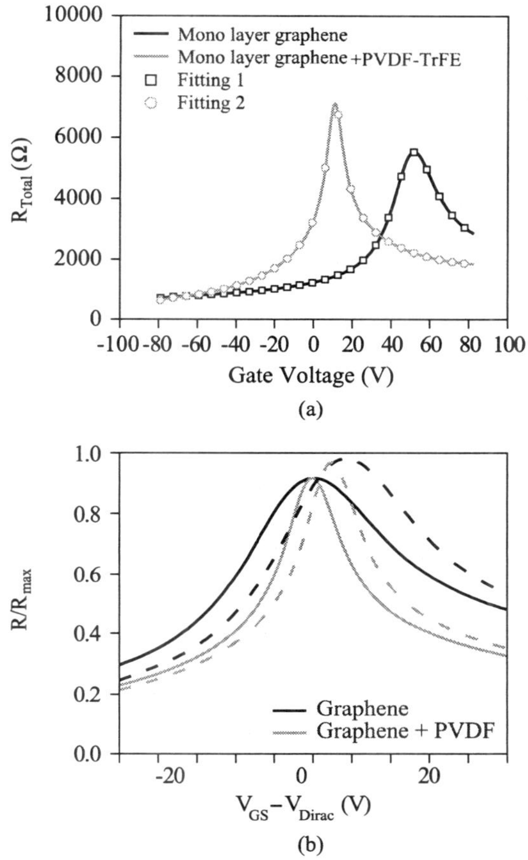

Figure 6.16 (a) Transfer characteristics of a GFET and (b) the normalized resistance of GFETs as a function of V_{GS} - V_{Dirac} before and after capping with P(VDF-TrFE) along with double sweeps of gate voltage

we observed all of the key device metrics in GFETs improved after the carbon-fluorine encapsulation. The transconductance in the GFETs increased by a factor of two, and the hysteresis was reduced, which is a phenomenon that had not been previously observed. Figure 6.16(a) shows the transfer characteristics of the GFET before and after capping with P(VDF-TrFE). Compared to the initial performance of the GFET device, the field-effect mobility increased from 1828 to 3241 cm^2/V-s, the residual carrier concentration decreased from 8.8 × 1011 to 4.7 × 1011 cm^{-2}, the Dirac voltage shifted toward zero from 50 to 9.5 V, and the on-off current ratio increased from 8 to 11, without change in contact resistance. The polymer piezoelectric P(VDF-TrFE) induces a strong dipole-dipole interaction at the interface of the electron-hole puddles in graphene. Notably, hysteresis in the GFET was reduced after capping it with P(VDF-TrFE), as shown in Figure 6.16(b). Such results indicate that the molecular encapsulation improves the interface characteristics of the GFETs by weakening the interaction between the charge carriers and the impurities (trap states).

The Raman spectra of the monolayer graphene before and after encapsulation with P(VDF-TrFE) are shown in Figure 6.17. The ratio of the intensities in the D and G peaks greatly increases from 0.2 to 1.9 after capping with P(VDF-TrFE) as a result of the structural change, via sp^{3-} hybridized dipole-interaction. Very likely, a new band at 2940 cm^{-1} was observed in the GFETs after encapsulating with P(VDF-TrFE), but this band was not observed in either bare graphene or P(VDF-TrFE) films. As discussed above, this result indicates that the monolayer graphene was chemically incorporated with fluoropolymer. We believe that these results strongly supports the origin of the interaction between fluoropolymer and graphene, which transforms the electrical characteristics of the GFETs. Though it is clear that the encapsulation with a class of fluoropolymer improves the quality of graphene-based devices for their higher on-off current ratio and symmetric electron-hole transport with improved mobility, further investigations are necessary to uncover the underlying mechanisms for such. A series of experimental and theoretical studies will further clarify the role of the fluoropolymer.

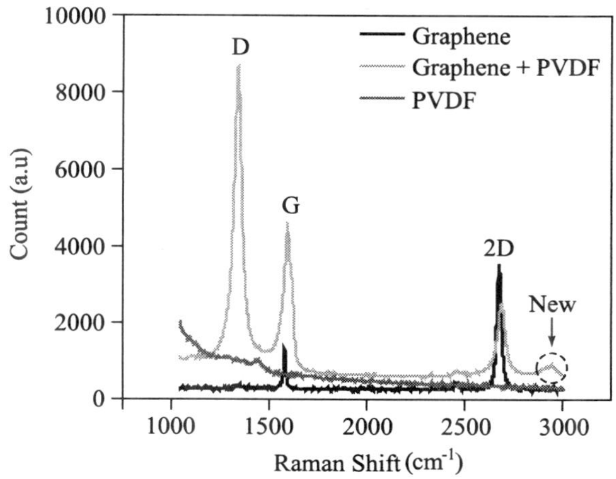

Figure 6.17 Raman spectra of monolayer graphene before and after capping with P(VDF-TrFE).

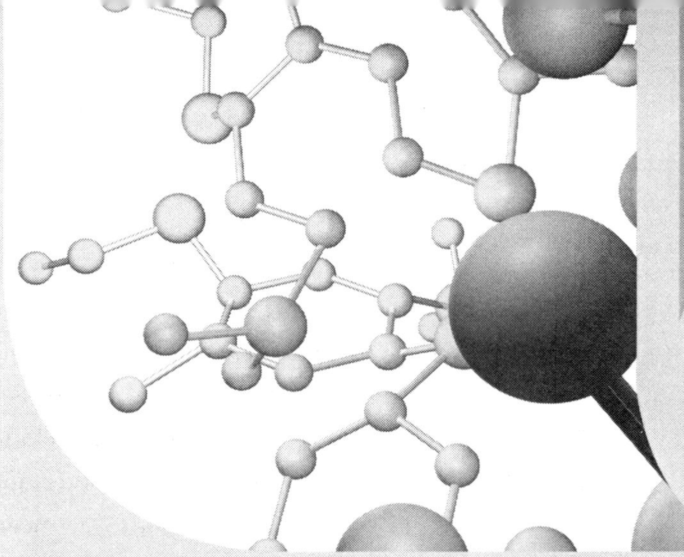

Chap 7

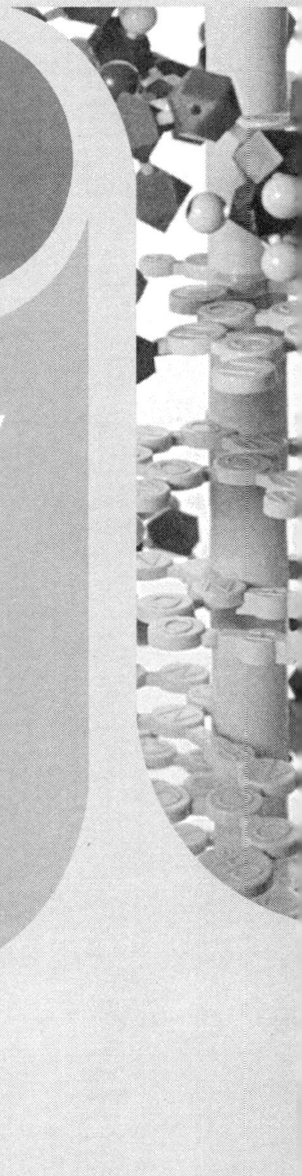

Summary

In this work we have realized the high performance graphene field-effect transistors on flexible plastic sheets. With the smoothened surface by applying and curing liquid PI, the surface roughness of the polymeric substrates could be significantly reduced so that the fabrication of novel devices are realized on those surfaces. The advanced technique using h-BN as the gate dielectric in PI films offers high electrical performance along with the first observation of the current saturation on plastic substrates. The embedded-gate device structure outperforms the conventional top-gate FET structure forseveral advantages: no need for the insulating seed layer prior to the gate dielectric deposition, dielectric scalability, and simpler fabrication process, which minimizes the chemical contamination to the graphene film during the device fabrication, which results in lower impurity density and higher electrical performance. The transistors are also robust against a variety of harsh conditions that are hazardous for conventional electronics, including immersion in liquids, and mechanical deformations; the devices remain fully functional with no significant degradation in their gate modulation. We have also investigated hybrid graphene/organic semiconductor FETs with α-6T and $F_{16}CuPc$. The electronic properties of graphene based FETs can be favorably tuned by capping with p-conjugated organic semiconductor molecules, particularly $F_{16}CuP$. The off-state current was reduced while the on-state current and mobility were either unaffected or increased. We have also shown that the electrical characteristics of graphene are favorably altered by capping with the fluoropolymer CYTOP®, Teflon-AF®, P(VDF-TrFE) . The on-off current ratio is improved and the Dirac voltage shifted toward zero. The residual carrier density no is reduced and the mobility increased by as much as a factor of two. We hypothesize that this alteration in electrical properties is a result of electronic polarization of the graphene by local C-F bonds or other dipoles in the fluoropolymer or semiconductor. The strength of this interaction is dependent on the annealing temperature, which influences the ordering of the polymer and consequently the local dipole field. The approach we described offers a way to transform the electrical characteristics of graphene making it potentially more useful for a wider range of electronic applications. We believe that this study motivates further research on flexible graphene field-effect transistors and also accelerate the realization of high-performance flexible nanoelectronics.

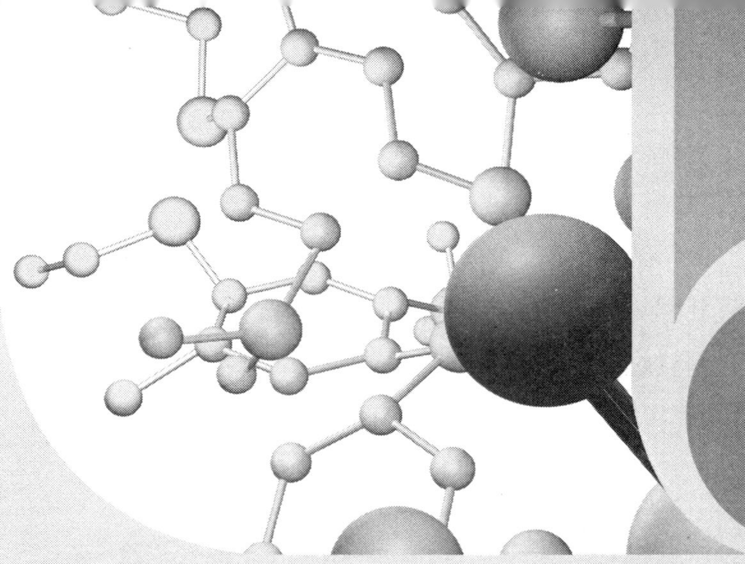

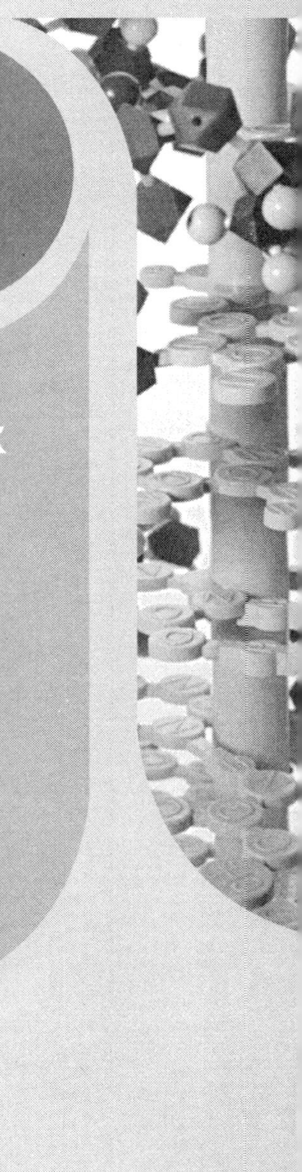

Chap 8

Appendix

A.1. Preparation of Double-Side Coated Polyimide Substrates

Step	Details
1	Clean polyimide substrates - Acetone rinse for 10 min - IPA rinse for 2 min - Blow dry with N_2 gun
2	Attach polyimide substrates to silicon carrier wafers for handling
3	Spin-coat liquid polyimide (PI-2574) on polyimide substrates - Spin rate : 2000 rpm - Ramp rate : 2000 rpm/sec - Time : 30 sec
4	Curing of polyimide substrates Equipment : NEYTECH furnace - Vent / load spin-coated samples / pump down - Turn on N_2 flow (remain on during the entire procedure) - Ramp up to 200°C at rate of 4°C/min - Hold for 30 min for soft-bake - Ramp up to 300°C at rate of 2.5°C/min - Hold for 60 min for cure - Cool down to 80°C at rate of −4°C/min - Turn off N_2 flow - Unload cured samples - Detach cured samples from silicon carrier wafers
5*	Double-side coating - Repeat step −2~4 for the other side of polyimide substrates
6**	Surface cleaning before lithography Equipment : Plasmatherm 790 - Oxygen plasma reactive ion etch for 10 sec(see Appendix A.3 for the details)

* This side is the top side of the substrate where devices will be patterned.

* The double-side curing process helps to cancel the built-in strain coming from the mismatch in the mechanical properties between the commercial polyimide film and the cured liquid layers while making the substrate more planar.

** This process helps lower the hydrophobicity of polyimide films to be used during lithography steps

A.2. Wet Transfer Process of Graphene Films

Step	Details
1	Chemical Vapor Deposition of Graphene using Cu foils
2	Attach Graphene/Cu foil samples on silicon carrier wafers
3	Spin-coat PMMA A4 on Graphene/Cu foil samples - Step-1. Spin-rate : 500 rpm, Ramp-rate : 1000 rpm/sec, Time : 5 sec - Step-2. Spin-rate : 4000 rpm, Ramp-rate : 2000 rpm/sec, Time : 40 sec - Step-3. Spin-rate : 0 rpm, Ramp-rate : 2000 rpm/sec, Time : 1 sec
4	Detach the spin-coated samples from silicon carrier wafers
5	Leave the samples in a desiccator overnight to remove residual solvents
6	Back-side graphene etching - Load the spin-coated samples upside-down in an RIE system - Oxygen plasma RIE for 1min to remove back-side graphene films (see Appendix A.3 for the details)
7	Cu wet-etch - Prepare clean, triple-rinsed beakers - Pour ammonia persulfate (APS-100) in each beaker - Load the sample in the beaker - Wait for 2 hours to complete etching Cu foils
8	Rinse samples with deionized (DI) water - Prepare three triple-rinsed beakers - Transfer one sample to the first beaker and wait for 30 min. - Transfer the sample to the second beaker and wait for 5 min. - Transfer the sample to the last beaker and wait for 5 min.
9	Transfer the sample to the target substrate
10	Repeat step-8 and step-9 for other samples
11	Dehydrate the samples overnight to remove residual moisture
12	Softbake samples at 180°C for 2~3min
13	Rinse samples with Acetone - Prepare a triple-rinsed beaker - Leave samples in the beaker for one day (Three cycles 6~8 hours each)
14	Visual inspection of transferred samples usingan optical microscope
15*	Raman spectroscopy to confirm the quality of films

* Raman spectroscopy is not available when polyimide films are used as target substrates. Users should prepare a reference sample (oxidized silicon wafer with 285-nm thick thermal oxide) for this process.

A.3. Oxygen Plasma Reactive Ion Etch of Graphene Films

Step	Details
1	Equipment : Plasmatherm 790 Place samples on a silicon carrier wafer Load samples with a carrier wafer at the center of the chamber
2	Pump (pressure : 100 mTorr, hold time : 10 sec)
3	Evacuation (pressure : 10 mTorr, hold time : 1 min)
4	Purge (pressure : 200 mTorr, hold time : 30 sec)
5	Evacuation (pressure : 10 mTorr, hold time : 30 sec)
6	Process step 1 O2flow rate : 10 sccm, pressure : 200 mTorr no RF power applied, hold time : 1 min (fixed)
7*	Process step 2 O_2 flow rate : 10 sccm, pressure : 200 mTorr RF power : 50 W, hold time : 40 sec
8	Evacuation (pressure : 10 mTorr, hold time : 30 sec)
9	Purge (pressure : 200 mTorr, hold time : 30 sec)
10	Evacuation (pressure : 10 mTorr, hold time : 30 sec)

* Etch for 40 sec for patterning graphene channel area with PMMA as a mask

A.4. Capture-Release Process for Exfoliated Hexagonal Boron Nitride Embedded in Polyimide

Step	Details
1	Prepare oxidized silicon wafer with 285 nm-thick thermal oxide
2	Rinse silicon substrates with Acetone and IPA for 2 min each
3	Spin-coat AZ-5209 (positive photo-resist) on silicon substrates - Spin-rate : 3000 rpm, Ramp-rate : 2000 rpm/sec, Time : 30 sec - Soft-bake at 95°C for 1 min
4	Optical lithography to define alignment marks on substrates Equipment : EVG-Aligner - Hard-contact, Exposure time : 2.5 sec - Develop samples using MIF-726 for 1 min
5	Ti/Au (2 nm/50 nm) deposition using e-beam evaporation
6	Liftoff of deposited metal films - Leave samples in Acetone for 1 hour and then rinse with IPA and DI water for 1 min each
7*	Surface cleaning of silicon wafers - Oxygen plasma RIE for 1 min
8	Exfoliate hexagonal boron nitride films - Exfoliate PolarTherm (PT-110) from Momentive Performance Materials Onto the prepared silicon substrates
9	Visual inspection of exfoliated samples using an optical microscope
10	Atomic force microscope to confirm the thickness of exfoliated films Equipment : VeecoVR D-5000 (in tapping mode)
11	Annealing of h-BN films Equipment : NEYTECH furnace - Temperature : 300°C under N_2 ambient - Time : 1 hour
12	Digitize locations for h-BN films - Capture optical images of h-BN films - Use a graphic digitizer to extract locations for each h-BN film - Design layouts using polygons
13	Spin-coat PMMA A4 on silicon substrates (see A.2 step-3) - Soft-bake at 180°C for 2 min for a thickness of 180 nm
14	E-beam lithography of embedded gates Equipment : Carl Zeiss SEM/EBL

	- Expose the desired patterns with a 20 kV electron beam at 240μC/cm^2 - Develop the pattern for 1min in 1 : 3 mixture of MIBK and isopropanol
15	Ti/Au (2 nm/50 nm) deposition using e-beam evaporation for gates
16	Liftoff of deposited metal films - Leave samples in Acetone for 1 hour and then rinse with IPA and DI water for 1 min each
17	Apply liquid polyimide - Follow A.1 step-2 to step-3
18**	Wet-etch thermal oxide (sacrificial layer) - Leave prepared substrates in BOE (1 : 6) for 4~6 hours - Pick up detached polyimide films - Rinse polyimide films with DI water - Dehydrate polyimide films overnight
19	Post annealing of the polyimide substrates with embedded patterns Equipment : NEYTECH furnace - Temperature : 300°C under N_2 ambient - Time : 1 hour
20	Wet-transfer of graphene films (see A.2 for details)
21	Spin-coat PMMA A4 and ESpacer on polyimide substrates - Spin-coat PMMA A4 and soft-bake (see A.4 step-13) - Spin-coat ESpacer Spin-rate : 3000 rpm, Ramp-rate : 2000 rpm/sec, Time : 60 sec
22	E-beam lithography of graphene channels Equipment : Carl Zeiss SEM/EBL - Expose the desired patterns with a 20 kV electron beam at 240μC/cm^2 - Develop the pattern for 1min in 1 : 3 mixture of MIBK and isopropanol
23	Oxygen plasma etch of graphene channels (see A.3 for details)
24	Spin-coat PMMA A4 and ESpacer on polyimide substrates (see A.4 step-21)
25	E-beam lithography of source/drain electrodes Equipment : Carl Zeiss SEM/EBL - Expose the desired patterns with a 20 kV electron beam at 240μC/cm^2 - Develop the pattern for 1min in 1 : 3 mixture of MIBK and isopropanol
26	Ti/Au (2 nm/50 nm) deposition using e-beam evaporation for source/drain
27	Liftoff of deposited metal films - Leave samples in Acetone for 1 hour and then rinse with IPA and DI water for 1 min each

* This improves the adhesion between h-BN films and substrates resulting in higher yield during the subsequent exfoliation process.

** Cured liquid polyimide films will be detached and this will work as a new substrate for device fabrication with fully embedded h-BN films.

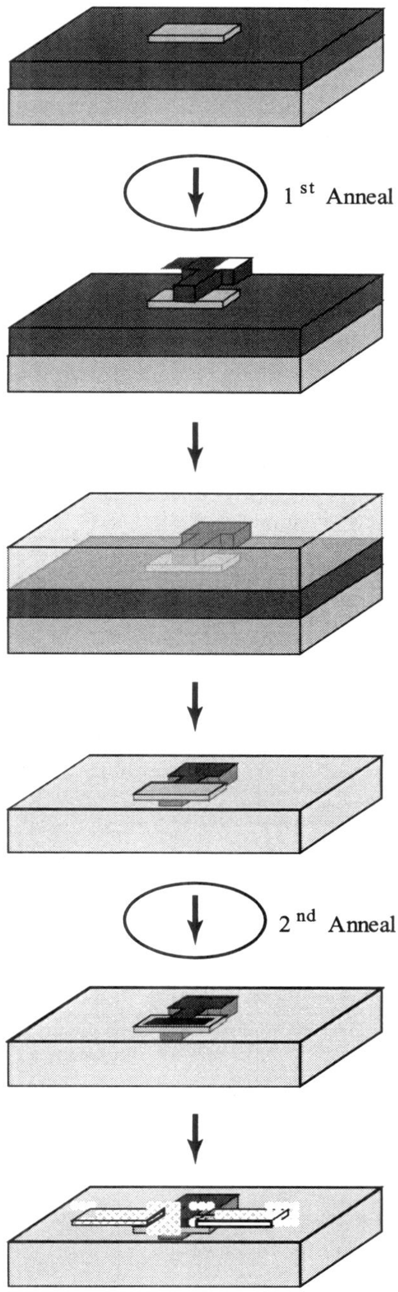

Figure A.1. Process steps for embedded-gate GFETs with h-BN films

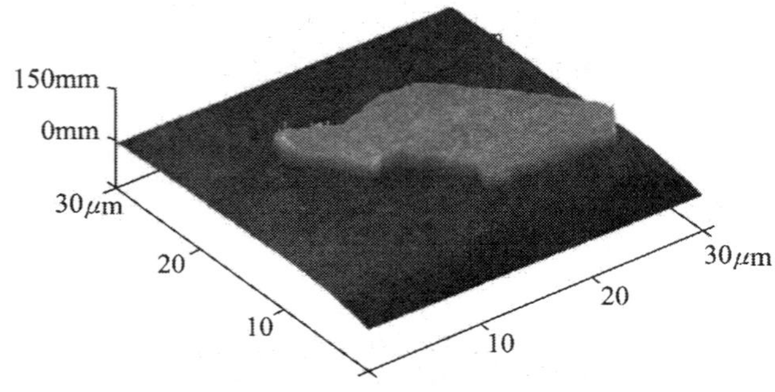

(a)

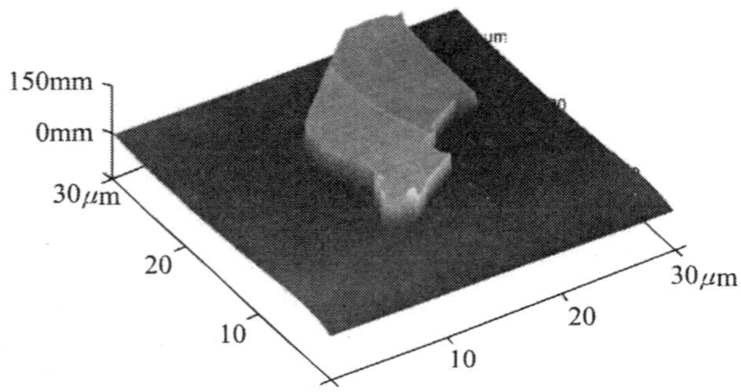

(b)

Figure A.2. Exfoliated h-BN films.

(a) AFM image captured as exfoliated,
(b) AFM image captured after annealing under nitrogen ambient

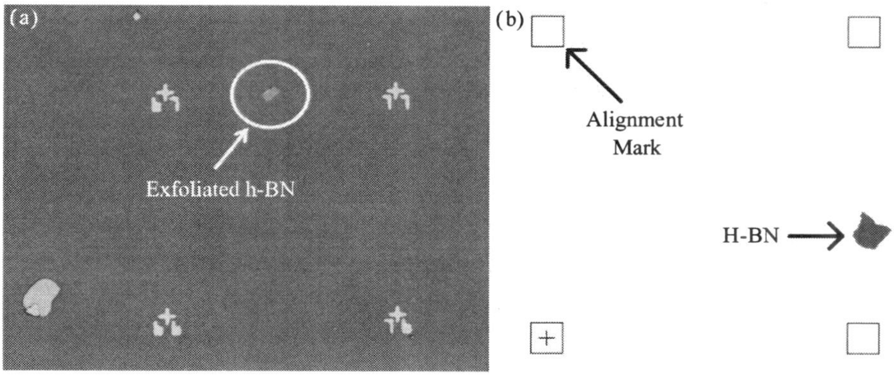

Figure A.3. Digitizing locations for h-BN films.

(a) Optical image of an exfoliated h-BN film,
(b) Designed layout. h-BN is drawn as a polygon.

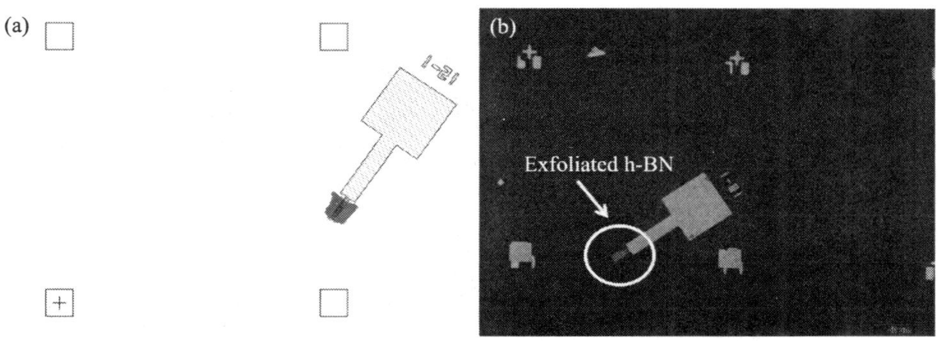

Figure A.4. Define embedded gate patterns on exfoliated h-BN films.

(a) Designed layout,
(b) Optical image captured after the liftoff of Ti/Au electrodes

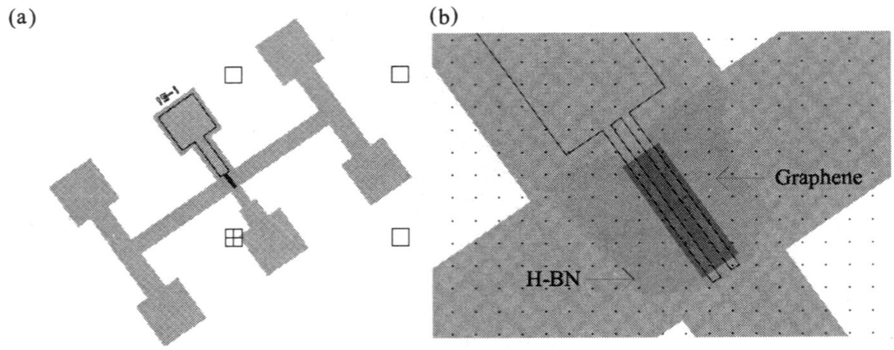

Figure A.5. Define graphene active layers

(a) Designed layout,
(b) Zoom-in picture of the channel area highlighting graphene channel and an h-BN dielectric film

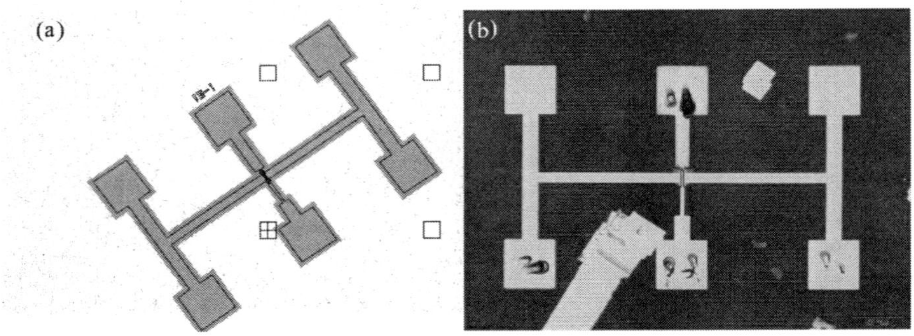

Figure A.6. Complete device structure with h-BN dielectric films.

(a) Designed layout forthe entire device structure including source/drain electrodes,
(b) Optical image captured after the liftoff of source/drain (Ti/Au) electrodes

A.5. Low Power Plasma Enhanced Chemical Vapor Deposition of N_2-Rich Si_3N_4

Step	Details
1	Equipment : Plasmatherm 790 Deposition temperature : 250 °C Place samples on a silicon carrier wafer Load samples with a carrier wafer at the center of the chamber
2	Pump (pressure : 20 mTorr, hold time : 10 sec)
3	Purge (pressure : 300 mTorr, hold time : 30 sec)
4	Evacuation (pressure : 20 mTorr, hold time : 2 min)
5	Purge (pressure : 300 mTorr, hold time : 30 sec)
6	Evacuation (pressure : 20 mTorr, hold time : 2 min)
7	Process step 1 NH_3 flow rate : 10 sccm, SiH_4 flow rate : 50 sccm, H_2 flow rate : 5 sccm, pressure : 500 mTorr, no RF power applied, hold time : 45 sec (fixed)
8*	Process step 2 NH_3 flow rate : 10 sccm, SiH_4 flow rate : 50 sccm, H_2 flow rate : 5 sccm, pressure : 500 mTorr, RF power : 10 W, hold time : variable
9	Evacuation (pressure : 10 mTorr, hold time : 30 sec)
10	Purge (pressure : 200 mTorr, hold time : 30 sec)
11	Evacuation (pressure : 10 mTorr, hold time : 30 sec)

* Si_3N_4 deposition rate is 53 A/min.

A.6. Selective Dry Etch of Plasma Enhanced Chemical Vapor Deposited Si3N4

Step	Details
1	Equipment : Plasmatherm 790 Place samples on a silicon carrier wafer Load samples with a carrier wafer at the center of the chamber
2	Pump (pressure : 100 mTorr, hold time : 10 sec)
3	Evacuation (pressure : 40 mTorr, hold time : 1 min)
4	Purge (pressure : 200 mTorr, hold time : 30 sec)
5	Evacuation (pressure : 40 mTorr, hold time : 30 sec)
6	Process step 1 Ar flow rate : 20 sccm, CF_4 flow rate : 50 sccm, pressure : 80 mTorr, no RF power applied, hold time : 1 min (fixed)
7*	Process step 2 Ar flow rate : 20 sccm, CF_4 flow rate : 50 sccm, pressure : 80 mTorr, RF 100 W, hold time : variable
8	Evacuation (pressure : 40 mTorr, hold time : 1 min)
9	Purge (pressure : 200 mTorr, hold time : 30 sec)
10	Evacuation (pressure : 40 mTorr, hold time : 1 min)

* The etch rate for Si_3N_4 is 114 nm/min.
* 950 PMMA A4 is used as a mask. The etch rate for PMMA is 88 nm/min. (selectivity of PMMA : Si_3N_4 = 1 : 1.29)

A.7. Multi-Finger Embedded-Gate Graphene Field Effect Transistor (MEGFET)

Step	Details
1	Prepare double-side coated polyimide substrates (see A.1 for details)
2	Spin-coat PMMA A4 and ESpacer on polyimide substrates (see A.4 step-21)
3	E-beam lithography of alignment marks Equipment : JEOL-6000 FSE - Expose the desired patterns with a 50 kV electron beam at 500μC/cm^2 - Develop the pattern for 1min in 1 : 3 mixture of MIBK and isopropanol
4	Ti/Au (2 nm/50 nm) deposition using e-beam evaporation for alignment marks
5	Liftoff of deposited metal films - Leave samples in Acetone for 1 hour and then rinse with IPA and DI water for 1 min each
6	Spin-coat PMMA A4 and ESpacer on polyimide substrates (see A.4 step-21)
7	E-beam lithography of gate patterns Equipment : JEOL-6000 FSE - Expose the desired patterns with a 50 kV electron beam at 500μC/cm^2 - Develop the pattern for 1min in 1 : 3 mixture of MIBK and isopropanol
8	Ti/Pd (2 nm/40 nm) deposition using e-beam evaporation for gate patterns
9	Liftoff of deposited metal films - Leave samples in Acetone for 1 hour and then rinse with IPA and DI water for 1 min each
10	Atomic layer deposition of Al_2O_3 - Deposition rate : 0.92 A/cycle at 250 °C - Cycles : 100 cycles
11	Spin-coat PMMA A4 and ESpacer on polyimide substrates (see A.4 step-21)
12	E-beam lithography of gate-dielectric patterns Equipment : JEOL-6000 FSE - Expose the desired patterns with a 50 kV electron beam at 500μC/cm^2 - Develop the pattern for 1min in 1 : 3 mixture of MIBK and isopropanol
13	Wet-etch ALD Al_2O_3 - Use a 1 : 3 mixed solution of H_3PO_4 : DI water - Etch rate : 0.3 nm/min

	- Etch patterned samples for > 35 min - Rinse samples with Acetone for 10 min - Rinse samples with IPA for 2 min - Rinse samples with DI water for 2 min
14	Wet-transfer of graphene films onto prepared substrates (see A.2 for details)
15	Spin-coat PMMA A4 and ESpacer on polyimide substrates (see A.4 step-21)
16	E-beam lithography of graphene channels Equipment : JEOL-6000 FSE - Expose the desired patterns with a 50 kV electron beam at 500μC/cm^2 - Develop the pattern for 1min in 1 : 3 mixture of MIBK and isopropanol
17	Oxygen plasma etch of graphene channels (see A.3 for details)
18	Spin-coat PMMA A4 and ESpacer on polyimide substrates (see A.4 step-21)
19	E-beam lithography of source/drain electrodes Equipment : JEOL-6000 FSE - Expose the desired patterns with a 50 kV electron beam at 500μC/cm^2 - Develop the pattern for 1min in 1 : 3 mixture of MIBK and isopropanol
20	Ti/Au (2 nm/50 nm) deposition using e-beam evaporation for source/drain
21	Liftoff of deposited metal films - Leave samples in Acetone for 1 hour and then rinse with IPA and DI water for 1 min each
22	Deposit a passivation layer of 30 nm-thick PECVD Si_3N_4 - Deposition rate : 5.3 nm/min - Deposition time : 6 min (see A.5 for details)
23	E-beam lithography of a Si_3N_4 passivation layer Equipment : JEOL-6000 FSE - Expose the desired patterns with a 50 kV electron beam at 500μC/cm^2 - Develop the pattern for 1min in 1 : 3 mixture of MIBK and isopropanol
24	Dry-etch PECVD Si_3N_4 passivation layer - Etch rate : 114 nm/min - Etch time : 20 sec (see A.6 for details) - Rinse samples with Acetone for 10 min - Rinse samples with IPA for 2 min

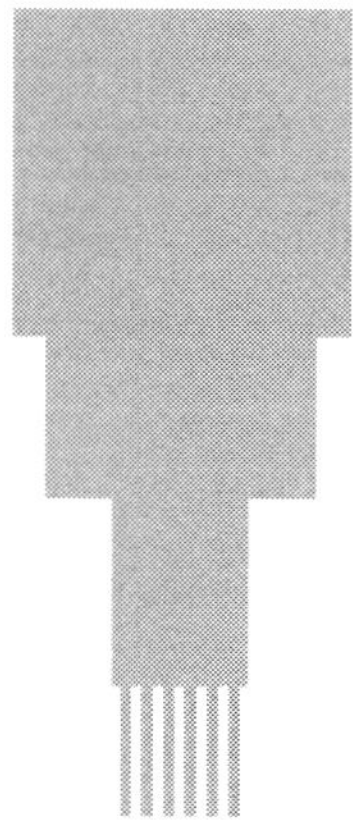

Figure A.7. Mask 1 for Multi-finger Embedded-gate Graphene Field-Effect Transistors to define gate fingers.

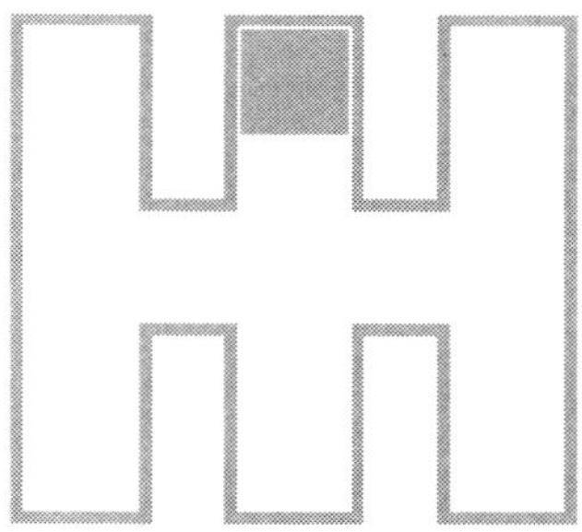

Figure A.8. Mask 2 for Multi-finger Embedded-gate Graphene Field-Effect Transistors to define ALD gate oxides.

Figure A.9. Mask 3 for Multi-finger Embedded-gate Graphene Field-Effect Transistors to define graphene active areas.

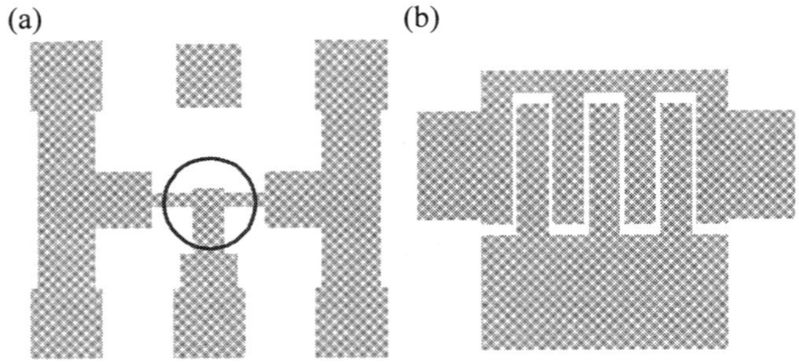

Figure A.10. (a) Mask4 for Multi-finger Embedded-gate Graphene Field-Effect Transistors to define source and drain electrodes. (b) Zoom-in picture over the channel area highlighted in red in (a).

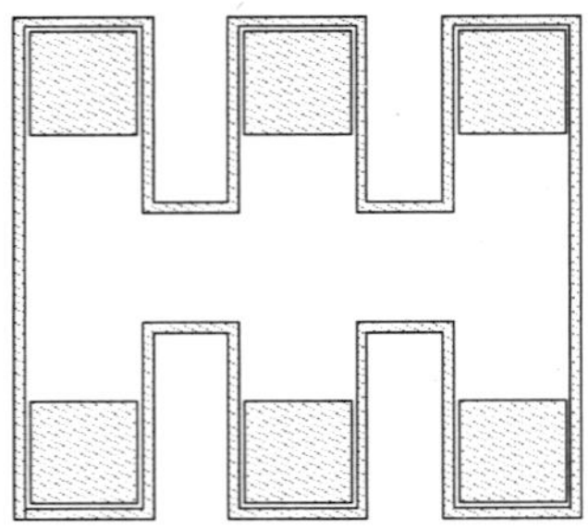

Figure A.11. Mask 5 for Multi-finger Embedded-gate Graphene Field-Effect Transistors to define the silicon nitride passivation layer.

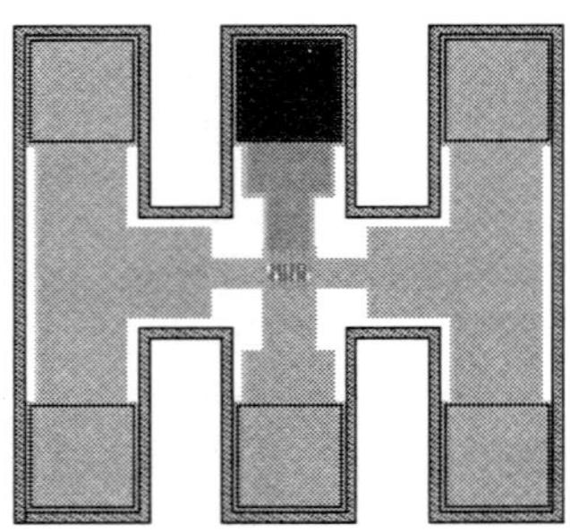

Figure A.12. Combined Layouts for Multi-finger Embedded-gate Graphene Field-Effect Transistors.

References

[1] Katsnelson, M. I. Graphene: carbon in two dimensions, Materialstoday, 2007, 10, 20

[2] Huang, X.; Qi, X.; Boey, F.; Zhang, H. Graphene-based composites, Chem. Soc. Rev. 2012, 41, 666.

[3] Castro Neto,A. H. Guinea, F.; Peres, N. M. R. Novoselov, K. S. .; and Geim, A. K. The Electronic properties of graphene, Rev. Mod, Phys. 2009, 81, 109.

[4] Wallace, P. R. The band Theory of Graphite, Phys. Rev. 1947, 71, 622.

[5] Novoselov, K. S. Geim, A. K. Morozov, S. V. Jiang, D.; Katsnelson, M. I. Grigorieva, I. V. Dubonos, S. V. Firsov, A. A. Two-dimensional gas of massless Dirac fermions in graphene, Nature, 2005, 438, 197.

[6] Bostwick, A.; Ohta, T.; Seyller, T.; Horn K.; Rotenberg, E. Quasiparticle dynamics in graphene, Nat. Physics, 2007, 3, 36.

[7] Son, D. T.Quantum critical point in graphene approached in the limit of infinitely strong Coulomb interaction, Phys. Rev. B. 2007, 75, 235423.

[8] Noveselov, K. S. Geim, A. K. Morozov, S. V. Jiang, D.; Zhang, Y.; Dubonos, S. V. Grigorieva, I. V. Firsov, A. A. Electric Field Effect in Atomically Thin Carbon Films, Science, 2004, 306, 666.

[9] Bolotin, K. I. Sikes, K. J. Jiang, Z.; Klima, M.; Fudenberg, G.; Hone, J.; Kim, P.; Stormer, H. L.Ultrahigh electron mobility in suspended graphene, Solid State Comm., 2008, 146, 351.

[10] Liao, L.; Lin, Y. –C. Bao, M.; Cheng, R.; Bai, J.; Liu, Y.; Qu, Y.; Wang, K. L.; Huang, Y.; Duan, X. High-speed graphene transistors with a self-aligned nanowire gate, Nature 2010, 467, 305.

[11] Konar, A.; Fang, T.; Jena, D. Effect of high-k gate dielectrics on charge transport in graphene-based field effect transistors, Phys. Rev. B. 2010, 82, 115452.

[12] Dean, C. R. Young, A. F.; Meric, I.; Lee, C.; Wang, L.; Sorgenfrei, S.;Watanabe, K.; Taniguchi, T.; Kim, P.; Shepard, K. L. Hone, J. Boron nitride substrates for high-quality graphene electronics, Nat. Nanotechnol., 2010, 5, 722.

[13] Giovannetti,G.; Khomyakov, P. A.; Brocks, G.; Kelly, P. J. van den Brink, J. Substrate-induced band gap in graphene on hexagonal boron nitride: Ab initio density functional calculations, Phys. Rev. B. 2007, 76, 073103.

[14] Fang, T.; Konar, A.; Xing, H.; Jena, D. High-field transport in two-dimensional graphene, Phys. Rev. B. 2011, 84, 125450.

[15] Schwierz, F. Graphene transistors, Nat. Nanotechnol., 2010, 5, 487.

[16] Zhou, S. Y. Gweon, G. –H.; Fedorov, A. V.; First, P. N.; De Heer, W. A.; Lee, D. –H. Guinea, F.; Castro Neto, A. H. Lanzara, A. Substrate-induced bandgap opening in epitaxial graphene, Nat. Mat. 2007, 6, 770.

[17] Ni, Z. H. Lu, T. Y. Y. H. Wang, Y. Y. Feng Y. O. Shen, Z. X. Uniaxial Strain on Graphene : Raman Spectroscopy Study and band-Gap Opening, ACS Nano, 2008, 2, 2301.

[18] Lin, Y. –M. Jenkins, K. A. Valdes-Garcia, A.; Small, J. P.; Farmer, D. B.; Avouris, P. Operation of Graphene Transistors at Gigahertz Frequencies, Nano Lett., 2009, 9, 422.

[19] Chen, J. –H. Jang, C.; Adam, S.; Fuhrer,M. S.; Williams, E. D.Charged-impurity scattering in graphene, Nat. Physics, 2008, 4, 377.

[20] Hwang, E. H. Adam, S.; Das Sarma, S. Carrier Transport in Two-Dimensional Graphene Layers, Phys. Rev. Lett., 2007, 98, 186806.

[21] Adam, S.; Hwang, E. H. Galitski, V. M. Das Sarma, S.;A self-consistent theory for graphene transport, PNAS, 2007, 104, 18392.

[22] Tworzydlo, J.; Trauzettel, B.; Titov, M.; Rycerz, A.; Beenakker, C. W. J. Sub-Poissonian Shot Noise in Graphene, Phys. Rev. Lett., 2006, 96, 246802.

[23] Du, X.; Skachko, I.; Barker, A.; Andrei, E. Y. Approaching ballistic transport in suspended graphene, Nat. Nanotechnol., 2008, 3, 491.

[24] Geim, a K.; Novoselov, K. S. The Rise of Graphene. Nat. Mater. 2007, 6, 183.

[25] Tan, Y.-W.; Zhang, Y.; Stormer, H. L.; Kim, P. Temperature Dependent Electron Transport in Graphene. The European Physical Journal Special Topics 2007, 148, 15.

[26] Schwierz, F. Graphene Transistors. Nat. Nanotechnol. 2010, 5, 487.

[27] Yang, X.; Liu, G.; Balandin, A. a; Mohanram, K. Triple-mode Single-transistor Graphene Amplifier and Its Applications. ACS nano 2010, 4, 5532.

[28] Link, C. Applications of Graphene Devices in RF Communications. 2013.48, 122.

[29] Tombros, N.; Veligura, A.; Junesch, J.; Jasper van den Berg, J.; Zomer, P. J.; Wojtaszek, M.; Vera Marun, I. J.; Jonkman,H. T.; Wees, B. J. van Large Yield Production of High Mobility Freely Suspended Graphene Electronic Devices on a Polydimethylglutarimide Based Organic Polymer. J. Appl. Phys. 2011, 109, 093702.

[30] Fallahazad, B.; Lee, K.; Kim, S.; Corbet, C.; Tutue, E.; Rd, B. (VDirac,TG). 2011, 062107, 1.

[31] Chen, J.-H.; Jang, C.; Xiao, S.; Ishigami, M.; Fuhrer, M. S. Intrinsic and Extrinsic Performance Limits of Graphene Devices on SiO2. Nat. nanotechnol. 2008, 3, 206.

[32] Perebeinos, V.; Avouris, P. Inelastic Scattering and Current Saturation in Graphene. Phys. Rev. B 2010, 81, 195442.

[33] Fratini, S.; Guinea, F. Substrate-limited Electron Dynamics in Graphene. Phys.Rev. B 2008, 77, 195415.

[34] Nathan, a.; Ahnood, a.; Cole, M. T.; Suzuki, Y.; Hiralal, P.; Bonaccorso, F.; Hasan, T.; Garcia-Gancedo, L.; Dyadyusha, a.; Haque, S. et al.Flexible Electronics: The Next Ubiquitous Platform. Proceedings of the IEEE 2012, 100, 1486.

[35] Dumitru, L. M.; Manoli, K.; Magliulo, M.; Sabbatini, L.; Palazzo, G.; Torsi, L. Plain Poly(acrylic Acid) Gated Organic Field-effect Transistors on a Flexible Substrate. ACS applied materials & interfaces 2013, 5, 10819.

[36] Briseno, a. L.; Tseng, R. J.; Ling, M.-M.; Falcao, E. H. L.; Yang, Y.; Wudl, F.; Bao, Z. High-Performance Organic Single-Crystal Transistors on Flexible Substrates. Adv.Mater. 2006, 18, 2320.

[37] Snow, E. S.; Campbell, P. M.; Ancona, M. G.; Novak, J. P. High-mobility Carbon-nanotube Thin-film Transistors on a Polymeric Substrate. Appl.Phys. Lett. 2005, 86, 033105.

[38] Cao, Q.; Kim, H.; Pimparkar, N.; Kulkarni, J. P.; Wang, C.; Shim, M.; Roy, K.; Alam, M. a; Rogers, J. a Medium-scale Carbon Nanotube Thin-film Integrated Circuits on Flexible Plastic Substrates. Nature 2008, 454, 495.

[39] Mack, S.; Meitl, M. a.; Baca, a. J.; Zhu, Z.-T.; Rogers, J. a. Mechanically Flexible Thin-film Transistors That Use Ultrathin Ribbons of Silicon Derived from Bulk Wafers. Appl.Phys. Lett. 2006, 88, 213101.

[40] Menard, E.; Nuzzo, R. G.; Rogers, J. a. High-speed Mechanically Flexible Single-crystal Silicon Thin-film Transistors on Plastic Substrates. IEEE Electron Dev. Lett. 2006, 27, 460.

[41] Celle, C.; Carella, A.; Mariolle, D.; Chevalier, N.; Rouvière, E.; Simonato, J.-P. Highly End-doped Silicon Nanowires for Field-effect Transistors on Flexible Substrates. Nanoscale 2010, 2, 677.

[42] Yoon, C.; Cho, G.; Kim, S. Electrical Characteristics of GaAs Nanowire-Based MESFETs on Flexible Plastics. IEEE Trans. on Electron Dev. 2011, 58, 1096.

[43] Kim, K. S.; Zhao, Y.; Jang, H.; Lee, S. Y.; Kim, J. M.; Kim, K. S.; Ahn, J.-H.; Kim, P.; Choi, J.-Y.; Hong, B. H. Large-scale Pattern Growth of Graphene Films for Stretchable Transparent Electrodes. Nature 2009, 457, 706.

[44] Lin, Y.-M.; Valdes-Garcia, A.; Han, S.-J.; Farmer, D. B.; Meric, I.; Sun, Y. Wu, Y.; Dimitrakopoulos, C.; Grill, A.; Avouris, P. et al.Wafer-scale Graphene Integrated Circuit. Science (New York, N.Y.) 2011, 332, 1294.

[45] Tao, L.; Lee, J.; Holt, M.; Akinwande, D. Wafer-scale Synthesis and Transfer of High Quality Monolayer Graphene for Nanoelectronics IEEE, 2011, 1.

[46] Tao, L.; Lee, J.; Holt, M.; Chou, H.; McDonnell, S. J.; Ferrer, D. a.; Babenco, M. G.; Wallace, R. M.; Banerjee, S. K.; Ruoff, R. S. et al.Uniform Wafer-Scale Chemical Vapor

Deposition of Graphene on Evaporated Cu (111) Film with Quality Comparable to Exfoliated Monolayer. The Journal of Physical Chemistry C 2012, 116, 24068.

[47] Garner, S. M.; He, M.; Lo, P.; Sung, C.; Liu, C.; Hsieh, Y.; Hsu, R.; Ding, J.; Hu, J.; Chan, Y. et al.Electrophoretic Displays Fabricated on Ultra-Slim Flexible Glass Substrates. J. Display Technol. 2012, 8, 590.

[48] Hoehla, S.; Garner, S.; Member, S.; Hohmann, M.; Kuhls, O.; Li, X.; Schindler, A.; Fruehauf, N. Requirements, A. S. Active Matrix Color-LCD on 75 m Thick Flexible Glass Substrates. J. Display Technol. 2012, 8, 309.

[49] Lee, S.; Chen, Y.; Liu, C. P.; Fan, T. J. Electrochemical Mechanical Polishing of Flexible Stainless Steel Substrate for Thin-Film Solar Cells. Int. J. Electrochem. Sci. 2013, 8, 6878.

[50] Afentakis, T.;Member, S.; Hatalis, M.; Member, S.; Voutsas, A. T.; Hartzell, J. Design and Fabrication of High-Performance Polycrystalline Silicon Thin-Film Transistor Circuits on Flexible Steel Foils. IEEE Trans. on Electron Dev. 2006, 53, 815.

[51] MacDonald, W. a.; Looney, M. K.; MacKerron, D.; Eveson, R.; Adam, R.; Hashimoto, K.; Rakos, K. Latest Advances in Substrates for Flexible Electronics. Journal of the Society for Information Display, 2007, 15, 1075.

[52] Wu, M.; Pangal, K.; Sturm, J. C.; Wagner, S. High Electron Mobility Polycrystalline Silicon Thin-film Transistors on Steel Foil Substrates. Appl. Phys. Lett. 1999, 75, 2244.

[53] Park, S. Y.; Kim, B. J.; Kim, K.; Kang, M. S.; Lim, K.-H.; Lee, T. Il; Myoung, J. M.; Baik, H. K.; Cho, J. H.; Kim, Y. S. Low-Temperature, Solution-processed and Alkali Metal Doped ZnO for High-performance Thin-film Transistors. Adv. Mater. (Deerfield Beach, Fla.) 2012, 24, 834.

[54] Lee, J.; Ha, T.; Parrish, K. N.; Member, S.; Chowdhury, S. F. High-Performance Current Saturating Graphene Field-Effect Transistor With Hexagonal Boron Nitride Dielectric on Flexible Polymeric Substrates. IEEE Trans.on Electron Dev, 2013, 34, 172.

[55] Petrone, N.; Meric, I.; Hone, J.; Shepard, K. L. Graphene Field-effect Transistors with Gigahertz-frequency Power Gain on Flexible Substrates. Nano Lett. 2013, 13, 121.

[56] Wong, W. S.; Salleo, A. Flexible Electronics Materials and Applications Springer.

[57] Hong, Y.; Chung, S.; Kattamis, A.; Cheng, I.-C.; Wagner, S. Technical Issues of Stainless Steel Foil Substrates for OLED Display Applications. Proc. of SPIE, 6655 (2007).

[58] Polyethylene Naphthalate (PEN) Material Information
http://www.goodfellow.com/E/Polyethylene-naphthalate.html.

[59] Polyimide (PI) Material Information http://www.goodfellow.com/E/Polyimide.html.

[60] Polyethylene Terephthalate (PET) Material Information
http://www.goodfellow.com/E/Polyethylene-terephthalate.html

[61] Tao, L.; Lee, J.; Akinwande, D. Nanofabrication down to 10 nm on a Plastic Substrate. Journal of Vacuum Science & Technology B: Microelectronics and Nanometer Structures 2011, 29, 06FG07.

[62] Hwang, E.; Adam, S.; Sarma, S. Carrier Transport in Two-Dimensional Graphene Layers. Phys. Rev. Lett. 2007, 98, 186806.

[63] Nomura, K.; MacDonald, a. H. Quantum Transport of Massless Dirac Fermions. Phys.Rev. Lett. 2007, 98, 076602.

[64] Chen, J.-H.; Ishigami, M.; Jang, C.; Hines, D. R.; Fuhrer, M. S.; Williams, E. D. Printed Graphene Circuits. Adv. Mater. 2007, 19, 3623.

[65] Xue, J.; Sanchez-Yamagishi, J.; Bulmash, D.; Jacquod, P.; Deshpande, A.; Watanabe, K.; Taniguchi, T.; Jarillo-Herrero, P.; LeRoy, B. J.Scanning Tunnelling Microscopy and Spectroscopy of Ultra-flat Graphene on Hexagonal Boron Nitride. Nat. Mater. 2011, 10, 282.

[66] Song, L.; Ci, L.; Lu, H.; Sorokin, P. B.; Jin, C.; Ni, J.; Kvashnin, A. G.; Kvashnin, D. G.; Lou, J.; Yakobson, B. I. et al. Large Scale Growth and Characterization of Atomic Hexagonal Boron Nitride Layers. Nano Lett. 2010, 10, 3209.

[67] Epitaxial, Q.; Bresnehan, M. S.; Hollander, M. J.; Wetherington, M.; Labella, M.; Trumbull, K. A.; Cavalero, R.; Snyder, D. W.; Robinson, J. A. Integration of Hexagonal Boron Nitride High-Performance Devices. ACS Nano, 2012, 6, 5234.

[68] Li, X.; Cai, W.; An, J.; Kim, S.; Nah, J.; Yang, D.; Piner, R.; Velamakanni, A.; Jung, I.; Tutuc, E. et al. Large-area Synthesis of High-quality and Uniform Graphene Films on Copper Foils. Science (New York, N.Y.) 2009, 324, 1312.

[69] Li, X.; Zhu, Y.; Cai, W.; Borysiak, M.; Han, B.; Chen, D.; Piner, R. D.; Colombo, L.; Ruoff, R. S. Transfer of Large-area Graphene Films for High-performance Transparent Conductive Electrodes. Nano Lett. 2009, 9, 4359.

[70] Pirkle, a.; Chan, J.; Venugopal, a.; Hinojos, D.; Magnuson, C. W.; McDonnell, S.; Colombo, L.; Vogel, E. M.; Ruoff, R. S.; Wallace, R. M. The Effect of Chemical Residues on the Physical and Electrical Properties of Chemical Vapor Deposited Graphene Transferred to SiO2. Appl. Phys.Lett.2011, 99, 122108.

[71] Lee, J.; Tao, L.; Hao, Y.; Ruoff, R. S.; Akinwande, D. Embedded-gate Graphene Transistors for High-mobility Detachable Flexible Nanoelectronics. Appl.Phys.s Lett. 2012, 100, 152104.

[72] Li, T.; Zhang, Z. Substrate-regulated Morphology of Graphene. J. Phys.D: Appl.Phys.2010, 43, 075303.

[73] Gao, W.; Huang, R. Effect of Surface Roughness on Adhesion of Graphene Membranes. J.Phys. D: Appl. Phys.s 2011, 44, 452001.

[74] Rutherglen, C.; Jain, D.; Burke, P. Nanotube Electronics for Radiofrequency Applications. Nat. Nanotechnol 2009, 4, 811.

[75] Liao, L.; Bai, J.; Cheng, R.; Lin, Y.-C.; Jiang, S.; Qu, Y.; Huang, Y.; Duan, X. Sub-100 Nm Channel Length Graphene Transistors. Nano Lett. 2010, 10, 3952.

[76] Parrish, K. N.; Akinwande, D. Impact of Contact Resistance on the Transconductance and Linearity of Graphene Transistors. Appl.Phys. Lett. 2011, 98, 183505.

[77] Pirkle, a.; Chan, J.; Venugopal, a.; Hinojos, D.; Magnuson, C. W.; McDonnell, S.; Colombo,

L.; Vogel, E. M.; Ruoff, R. S.; Wallace, R. M. The Effect of Chemical Residues on the Physical and Electrical Properties of Chemical Vapor Deposited Graphene Transferred to SiO2. Appl.Phys. Lett. 2011, 99, 122108.

[78] Hsu, A.; Wang, H.; Kim, K. K.; Kong, J.; Palacios, T. Impact of Graphene Interface Quality on Contact Resistance and RF Device Performance. IEEE Electron Dev. Lett. 2011, 32, 1008.

[79] Giovannetti, G.; Khomyakov, P.; Brocks, G.; Karpan, V.; Brink, J. van den; Kelly, P. Doping Graphene with Metal Contacts. Phys. Rev. Lett.2008, 101, 026803.

[80] Nagashio, K.; Nishimura, T.; Kita, K.; Toriumi, A. Metal / Graphene Contact as a Performance Killer of Ultra-high Mobility Graphene - Analysis of Intrinsic Mobility and Contact Resistance -. IEDM, 2009, 1.

[81] Xia, F.; Perebeinos, V.; Lin, Y.; Wu, Y.; Avouris, P. The Origins and Limits of Metal-graphene Junction Resistance. Nat. Nanotechnol. 2011, 6, 179.

[82] Huard, B.; Stander, N.; Sulpizio, J.; Goldhaber-Gordon, D. Evidence of the Role of Contacts on the Observed Electron-hole Asymmetry in Graphene. Phys. Rev. B 2008, 78, 121402.

[83] Lee, J.; Tao, L.; Hao, Y.; Ruoff, R. S.; Akinwande, D. Embedded-gate Graphene Transistors for High-mobility Detachable Flexible Nanoelectronics. Appl. Phys. Lett. 2012, 100, 152104.

[84] Lee, J.; Tao, L.; Parrish, K. N.; Hao, Y.; Ruoff, R. S.; Akinwande, D. Multi-finger Flexible Graphene Field Effect Transistors with High Bendability. Appl. Phys. Lett. 2012, 101, 252109.

[85] Smith, J. T.; Franklin, A. D.; Farmer, D. B.; Dimitrakopoulos, C. D. Reducing Contact Resistance in Graphene Devices Through Contact Area Patterning. ACS nano 2013, 7, 3661.

[86] Han, S.-J.; Valdes-Garcia, A.; Bol, A. a.; Franklin, A. D.; Farmer, D.; Kratschmer, E.; Jenkins, K. a.; Haensch, W. Graphene Technology with inverted-T Gate and RF Passives on 200 Mm Platform. IEDM 2011, 2.2.1–2.2.4.

[87] Zhu, W.; Neumayer, D.; Perebeinos, V.; Avouris, P. Silicon Nitride Gate Dielectrics and Band Gap Engineering in Graphene Layers. Nano Lett. 2010, 10, 3572.

[88] Ha, T.-J.; Lee, J.; Chowdhury, S. F.; Akinwande, D.; Rossky, P. J.; Dodabalapur, A. Transformation of the Electrical Characteristics of Graphene Field-effect Transistors with Fluoropolymer. ACS applied materials & interfaces 2013, 5, 16.

[89] Kim, S.; Nah, J.; Jo, I.; Shahrjerdi, D.; Colombo, L.; Yao, Z.; Tutuc, E.; Banerjee, S. K. Realization of a High Mobility Dual-gated Graphene Field-effect Transistor with Al2O3 Dielectric. Appl.Phys.Lett. 2009, 94, 062107.

[90] Lee, J.; Ha, T.-J.; Li, H.; Parrish, K. N.; Holt, M.; Dodabalapur, A.; Ruoff, R. S.; Akinwande, D. 25 GHz Embedded-Gate Graphene Transistors with High-K Dielectrics on Extremely Flexible Plastic Sheets. ACS nano 2013, 7, 7744.

[91] Lee, B.; Chen, Y.; Duerr, F.; Mastrogiovanni, D.; Garfunkel, E.; Andrei, E. Y.; Podzorov, V. Modification of Electronic Properties of Graphene with Self-assembled Monolayers. Nano

Lett. 2010, 10, 2427.

[92] Lafkioti, M.; Krauss, B.; Lohmann, T.; Zschieschang, U.; Klauk, H.; Klitzing, K. V; Smet, J. H. Graphene on a Hydrophobic Substrate: Doping Reduction and Hysteresis Suppression Under Ambient Conditions. Nano Lett. 2010, 10, 1149.

[93] Ha, T.; Lee, J.; Member, S.; Member, D. A. The Restorative Effect of Fluoropolymer Coating on Electrical Characteristics of Graphene Field-Effect Transistors. IEEE Electron Dev. Lett. 2013, 34, 559.

[94] DiBenedetto, S. a.; Facchetti, A.; Ratner, M. a.; Marks, T. J. Molecular Self-Assembled Monolayers and Multilayers for Organic and Unconventional Inorganic Thin-Film Transistor Applications. Adv. Mater. 2009, 21, 1407.

[95] Ram, M. E.; Member, S.; Parrish, K. N.; Chowdhury, S. F.; Magnuson, C. W.; Movva, H. C. P.; Ruoff, R. S.; Banerjee, S. K. Three-Gigahertz Graphene Frequency Doubler on Quartz Operating Beyond the Transit Frequency. IEEE Trans. on Nanotechnol. 2012, 11, 877.

[96] Nayfeh, O. M. Graphene Transistors on Mechanically Flexible Polyimide Incorporating Atomic-Layer-Deposited Gate Dielectric. IEEE Electron Dev. Lett. 2011, 32, 1349.

[97] Sire, C.; Ardiaca, F.; Lepilliet, S.; Seo, J.-W. T.; Hersam, M. C.; Dambrine, G.; Happy, H.; Derycke, V. Flexible Gigahertz Transistors Derived from Solution-based Single-layer Graphene. Nano Lett. 2012, 12, 1184.

[98] Parrish, K. N.; Akinwande, D. An Exactly Solvable Model for the Graphene Transistor in the Quantum Capacitance Limit. Appl. Phys. Lett. 2012, 101, 053501.

[99] Schwierz, G. Graphene transistors: status, prospects and problem, Proc. of the IEEE, 2013, 101, 1567.

[100] Wang, H.; Hsu, A.; Kim, K. K.; Kong, J.; Palacios, T. Gigahertz Ambipolar Frequency Multiplier Based on CVD Graphene.IEDM, 2010, 1.

[101] Parrish, K. N.; Akinwande, D. Even-odd Symmetry and the Conversion Efficiency of Ideal and Practical Graphene Transistor Frequency Multipliers. Appl. Phys. Lett. 2011, 99, 223512.

[102] Liao, L.; Lin, Y.-C.; Bao, M.; Cheng, R.; Bai, J.; Liu, Y.; Qu, Y.; Wang, K. L.; Huang, Y.; Duan, X. High-speed Graphene Transistors with a Self-aligned Nanowire Gate. Nature 2010, 467, 305.

[103] Lin, Y.; Dimitrakopoulos, C.; Jenkins, K. A.; Farmer, D. B.; Chiu, H.; Grill, A.; Avouris, P. 100-GHz transistors fromwafer-scale epitaxial graphene, Science, 2010, 327, 662.

[104] Cha, J.; Cha, J.; Lee, S.; Member, S. Uncertainty Analysis of Two-Step and Three-Step Methods for Deembedding On-Wafer RF Transistor Measurements. IEEE Electron Dev. Lett. 2008, 55, 2195–2201.

[105] Farmer, D. B.; Valdes-Garcia, A.; Dimitrakopoulos, C.; Avouris, P. Impact of Gate Resistance in Graphene Radio Frequency Transistors. Appl. Phys. Lett. 2012, 101, 143503.

[106] Kim, D.; Brar, B.; Alamo, J. A. f_T = 688GHz and fmax=800 GHz in Lg=40nm

In0.7Ga0.3As MHEMTs with g m _ Max > 2 . 7 mS / Mm. IEDM 2011, 1.

[107] Petrone, N.; Meric, I.; Hone, J.; Shepard, K. L. Graphene Field-effect Transistors with Gigahertz-frequency Power Gain on Flexible Substrates. Nano Lett. 2013, 13, 121.

[108] Lee, T. H. The Designof CMOS Radio-Frequency Integrated Circuits 2nd ed.; Cambridge University Press: Cambridge, UK, 2004.

[109] Wu, Y.; Jenkins, K. a; Valdes-Garcia, A.; Farmer, D. B.; Zhu, Y.; Bol, A. a; Dimitrakopoulos, C.; Zhu, W.; Xia, F.; Avouris, P. et al. State-of-the-art Graphene High-frequency Electronics. Nano Lett. 2012, 12, 3062.

[110] Lee, C.; Wei, X.; Kysar, J. W.; Hone, J. Measurement of the Elastic Properties and Intrinsic Strength of Monolayer Graphene. Science (New York, N.Y.) 2008, 321, 385.

[111] Li, T.; Huang, Z.; Suo, Z.; Lacour, S. P.; Wagner, S. Stretchability of Thin Metal Films on Elastomer Substrates. Appl. Phys. Lett. 2004, 85, 3435.

[112] Kang, J.; Kim, H.; Kim, K. S.; Lee, S.-K.; Bae, S.; Ahn, J.-H.; Kim, Y.-J.; Choi, J.-B.; Hong, B. H. High-performance Graphene-based Transparent Flexible Heaters. Nano Lett. 2011, 11, 5154

[113] Scarpa, F.; Adhikari, S.; Gil, A. J.; Remillat, C. The bending of single layer graphene sheets: the lattice versus continuum approach, Nanotechnology, 2010, 21, 125702.

[114] Chang, H.-Y.; Yang, S.; Lee, J.; Tao, L.; Hwang, W.-S.; Jena, D.; Lu, N.; Akinwande, D. High-performance, Highly Bendable MoS2 Transistors with High-k Dielectrics for Flexible Low-power Systems. ACS nano 2013, 7, 5446.

[115] Crone, B.; Dodabalapur, A. Lin, Y. –Y. Filas, R. W. Bao, Z.; LaDuca, A.; Sarpeshkar, R.; Katz, H. E. Li, W.Large-Scale complementary integrated circuits based on organic transistors, Nature, 1999, 403, 521.

[116] Sahin, H.; Topsakal, M.; Ciraci, S. Structures of fluorinated graphene and their signatures, Phys. Rev. B. 2011, 83, 115432.

[117] Robinson, J. T. Burgess, J. S.; Junkermeier, C. E.; Badescu, S. C.; Reinecke, T. L.; Perkins, F. K.; Zalautdniov, M. K.; Baldwin, J. W.; Culbertson, J. C.; Sheehan, P. E.; Snow, E. S.Properties of Fluorinated Graphene Film, Nano. Lett. 2010, 10, 3001.

[118] Li, X.; Zhu, Y.;Cai, W.; Borysiak, M.; Han, B.; Chen,D.; Piner, R. D.; Colombo, L.; Ruoff, R. S.Transfer of Large-Area Graphene Films for High-Performance Transparent Conductive Electrodes, Nano Lett., 2009, 9, 4359.

[119] Lin, Y. -M. Chiu, H. -Y. Jenkins, K. A. Farmer, D. B. Avouris, P.; Valdes-Garcia, A. Dual Gate Graphene FETs with f_T of 50 GHz, IEEE Electron Dev.Lett. 2010, 31, 68.

[120] Chen, F.; Xia, J.;Ferry, D. K. Tao, N. Dielectric screening enhanced performance in graphene FET, Nano Lett., 2009, 9, 2571.

Tae-Jun Ha received his Ph.D. in Electrical and Computer Engineering from The University of Texas at Austin in 2012. Previously, he graduated with a B.S and a M.S degree in Electrical Engineering and Computer Science from Seoul National University in 2007 and 2009, respectively. Between 2012 and 2013 he was with the nanomanufacturing systems for mobile computing and mobile energy technologies (NASCENT) at the University of Texas at Austin and University of California at Berkeley. Since 2014, he is with Kwangwoon University. His current research includes thin-film transistors based on advanced materials and nano-scale fabrication process, system-leveled human-interactive sensors and charge transport physics.

Jong-Ho Lee received a Ph.D in Electrical and Computer Engineering from the University of Texas at Austin in May 2014. Previously, he graduated with a B.S degree in Electrical Engineering and a M.E. degree in Electrical Engineering from Korea University in 2006 and 2008, respectively. His Ph.D thesis focused on the fabrication of Graphene Field-Effect Transistors for flexible electronics. He has published diverse journal papers as author and co-author for 2D materials including Graphene and MoS2; their characteristics, device fabrication, and applications. In September 2014, he joined Samaung Electronics in South Korea where he develops and characterizes next node noble CMOS logic devices.

Graphene: 정가 15,000원
charge transport and flexible nanoelectronics

인 쇄 2016년 01월 02일 초판 1쇄
발 행 2016년 01월 10일 초판 1쇄
저 자 Tae-Jun Ha · Jong-Ho Lee
발행인 정우용
발행처 도서출판 동화기술
경기도 파주시 광인사길 201(문발동, 파주출판도시)
Tel (031)955-4211~6 donghwae@kornet.net
Fax (031)955-4217 www.donghwapub.co.kr
(등록) 1977년 12월 19일/9-16호

ISBN 978-89-425-9048-3